CHINA 2020

SHARING RISING INCOMES

CHINA 2020 SERIES

China 2020:
Development Challenges in the New Century

Clear Water, Blue Skies:
China's Environment in the New Century

At China's Table:
Food Security Options

Financing Health Care:
Issues and Options for China

Sharing Rising Incomes:
Disparities in China

Old Age Security:
Pension Reform in China

China Engaged:
Integration with the Global Economy

THE WORLD BANK
WASHINGTON D.C.

SHARING RISING INCOMES

DISPARITIES IN CHINA

First printing September 1997

The World Bank does not guarantee the accuracy of the data included in this publication and accepts no responsibility whatsoever for any consequence of their use. The boundaries, colors, denominations, and other information shown on any map in this volume do not imply on the part of the World Bank Group any judgment on the legal status of any territory or the endorsement or acceptance of such boundaries.

Cover photograph by Claus Meyer/Black Star.

Cover insets (from left to right) by Vince Streano/Aristock, Inc.; Dennis Cox/China Stock; Serge Attal/Gamma Liaison; Dennis Cox/China Stock; Joe Carini/Pacific Stock; Erica Lansner/Black Star.

ISBN: 0-8213-4075-1

Contents

This report uses *Hong Kong* when referring to the Hong Kong Special Administrative Region, People's Republic of China.

Acknowledgments

This report was written by Tamar Manuelyan Atinc based on inputs from many individuals. Discussions with Chinese officials during a July 1996 mission were particularly helpful in identifying the main challenges facing the authorities and understanding the institutional context in which policies affecting income distribution evolve. The mission, comprising Tamar Manuelyan Atinc, Valerie Charles, Albert Keidel, and Julia Li, is particularly thankful for the assistance of the State Planning Commission's Spatial and Regional Planning Department. The contributions of Chen Xuan Qing, Chen Xiang, and Yan Pangui are gratefully acknowledged.

The study also could not have been carried out without the help of the State Statistical Bureau. Information provided by the urban and rural household survey team and background papers prepared by the bureau's Research Institute were essential ingredients for the report.

The report also benefited from analytical inputs, some in the form of background papers, from many scholars who have shown keen interest in income inequality. Within the World Bank these included Shaohua Chen, Yuri Dikhanov, Francisco Ferreira, Marcel Fratzscher, Shaikh Hossain, Aart Kraay, Martin Ravallion, Christine Wong, Colin Xu, Xiaoqing Yu, Tao Zhang, and Heng-Fu Zou. Outside the Bank, Robin Burgess (London School of Economics/STICERD) and Calla Wiemer (University of Hawaii) made valuable contributions.

Valuable comments were also received from peer reviewers, including Stephen Howes (World Bank), Nora Lustig (Brookings Institution), Carl Riskin (Columbia University), and Lyn Squire (World Bank). Other World Bank staff were also generous with their time and advice, including Vinod Ahuja, Liang Li, Natalie Lichtenstein, Andrew Mason, Richard Newfarmer, Vikram Nehru, and Alan Piazza. Bonita Brindley provided valuable advice on writing. Klaus Rohland and Nicholas Hope provided strategic guidance and able management.

The report was edited by Meta de Coquereaumont and Paul Holtz, laid out by Damon Iacovelli and Laurel Morais, and designed by Kim Bieler, all with the American Writing Division of Communications Development Incorporated.

China's GDP per capita has grown at a remarkable 8.2 percent a year since economic reforms started in 1978. Market incentives have diversified employment, increased factor mobility, and boosted returns to land and human capital. Overall, a staggering 200 million people have been lifted out of poverty.

But the benefits of growth are unevenly distributed. People with schooling, mobility, and good land have been able to take advantage of the new market opportunities, helping to spur growth. But government policies, or their absence, are heightening inequalities. Social policies favor urban areas. Economic policies favor the coast. Access to education, health care, and employment opportunities remains unequal or has become more so. And gender disparities in the marketplace may be more pronounced.

Elsewhere, high inequality has depressed growth, undermined poverty alleviation, and contributed to social tension. China's income inequality, similar to that in the United States, remains moderate by international standards. But even though it may continue to rise as the country's transition unfolds, increased inequality need not undermine growth or social harmony—provided it is accompanied by broadly based growth, equal access to opportunities, and protection for the poor and vulnerable. The challenge for the Chinese government is to extend the benefits of growth to all members of society.

Overview

China's income distribution has become increasingly unequal since reforms started in 1978. The Gini coefficient (a common measure of income inequality), a low 28.8 in 1981, reached 38.8 in 1995. A change of this magnitude is highly unusual and signals deep structural transformation in the distribution of assets and their returns.

Inequality has risen in large part because China has begun to harness the enormous potential of its people, suppressed during the first three decades of Communist rule. At the height of egalitarianism individual remuneration barely reflected productivity. In 1978 the government introduced individual incentives and market forces that immediately began to increase returns to capital and land, diversify employment, and increase factor mobility. Not surprisingly, the benefits of growth were distributed unevenly, accruing to those most able to take advantage of rising opportunities—the educated and the enterprising, the mobile, and those with high-quality land. To some degree inequality was necessary for the rapid growth that followed the adoption of reforms. But government policies, or their absence, are exacerbating inequalities. Social policies favor urban over rural areas, economic policies favor the coast over the interior, and access to education, health care, and labor mobility remains unequal or has become more so. And the price of admission to a more affluent society appears to be higher for women than for men.

Should China's policymakers be concerned about the increasing polarization of incomes? Elsewhere, high inequality has impeded growth, undermined poverty alleviation, and contributed to social tension. China's income inequality is still moderate. The benefits of growth have been unevenly distributed, but they have reached the poor. Moreover, much of the increase in inequality reflects a welcome adjustment to an incentive and remuneration structure more typical of market economies. But if not moderated, some aspects of China's inequality may imperil future growth and stability.

Social tension can result when the benefits of growth accrue unequally to easily identifiable groups—for example, geographic and urban-rural imbalances, inequalities between ethnic groups, and gender disparities—even if these are not major factors in explaining overall income inequality. If richer groups enjoy consistently higher growth, simmering social tensions can become politically destabilizing and ultimately derail growth and poverty reduction. Social and economic progress can also be damaged by rising inequalities in opportunities. Experience elsewhere suggests that inequalities in access to basic health and education typically accompany higher income inequality and can intensify its negative effects on society. Policymakers in China need to manage the widening gap between rural and urban areas, the growing disparities between the coast and the interior, and the increasing inequality across income groups in access to opportunities for self-improvement.

Progress and problems

Although China's income inequality has risen rapidly, it has not yet pushed the country into the ranks of the notoriously unequal. China's Gini coefficient is now similar to that of the United States and close to the East Asian average—substantially higher than in Eastern Europe but much lower than in Sub-Saharan Africa and Latin America (table 1).

Moreover, China's spectacular growth has been accompanied by substantial gains in poverty reduction. Since the start of reforms in 1978, China has lifted some 200 million people out of absolute poverty. But progress has been uneven. Most of the poverty reduction occurred in the early part of reforms, when the household responsibility system was introduced in rural areas. But in the mid-1980s and early 1990s poverty levels stagnated despite steady gains in per capita GDP. Since 1992 renewed momentum has decreased the number of poor, and by the end of 1995 less than 6 percent of the population had incomes below the absolute poverty line.

Growth in rural incomes has transformed poverty statistics. Per capita GDP growth did not always increase personal incomes, but when it boosted rural incomes, poverty declined. Without rural income growth, the number of absolute poor in China would have increased by more than 100 million between 1981 and 1995 because of adverse distributional changes. Instead, the ranks of the poor fell by more than 150 million.

But there is no room for complacency. Reforms have not reduced the large welfare differences between rural and urban households; on the contrary, these have increased. Policies favoring the coast have reinforced the region's natural endowments, widening the gulf between coastal and interior provinces. Market forces have raised productivity, but labor markets remain segmented. And if the marketplace alone is left to dictate

TABLE 1
China's inequality puts it in the middle of the pack internationally
(Gini coefficient)

Region or country	1980s	1990s
Eastern Europe	25.0	28.9
China[a]	**28.8**	**38.8**
High-income countries	33.2	33.8
South Asia	35.0	31.8
East Asia and the Pacific	38.7	38.1
Middle East and North Africa	40.5	38.0
Sub-Saharan Africa	43.5	47.0
Latin America and the Caribbean	49.8	49.3

a. Data are for 1981 and 1995.
Source: Deininger and Squire 1996; Ahuja and others 1997.

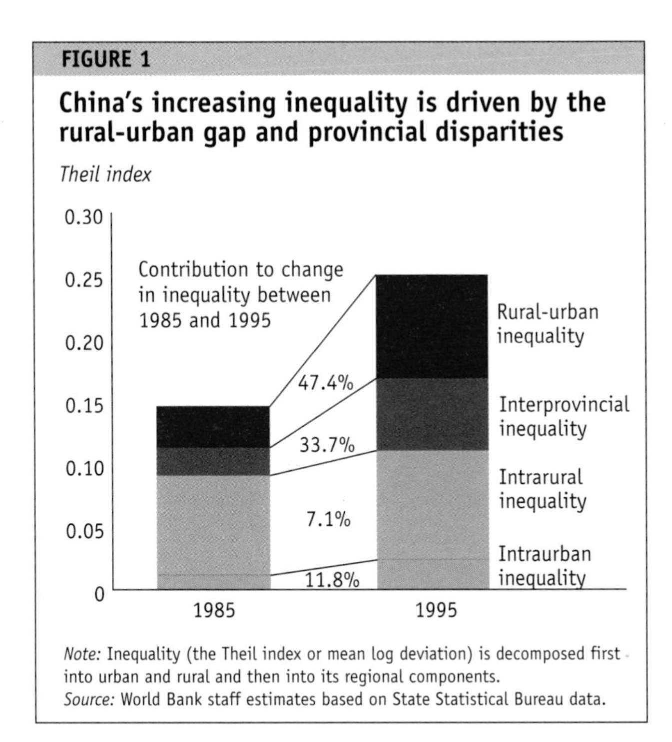

FIGURE 1

China's increasing inequality is driven by the rural-urban gap and provincial disparities

Note: Inequality (the Theil index or mean log deviation) is decomposed first into urban and rural and then into its regional components.
Source: World Bank staff estimates based on State Statistical Bureau data.

social conditions, the quality of China's human resources may become more and more uneven, creating and isolating winners and losers based on education, assets, and, increasingly, gender.

The rural-urban divide is increasing

China's urban dwellers enjoy a considerably higher standard of living than their rural counterparts. Rural incomes grew rapidly in the early period of reforms but in 1985 began to trail the increases in urban incomes, a trend reversed only in 1995. According to official data, the rural-urban income gap explained one-third of total inequality in 1995 and one-half of the increase in inequality since 1985 (figure 1). Internationally, the urban-rural income ratio rarely exceeds 2.0—as it does in China—and in most countries it is below 1.5. But even China's high ratio fails to capture the full extent of disparities in living standards between city dwellers and rural residents. An elaborate set of publicly provided services—housing, pensions, health, education, and other entitlements—augment urban incomes by an average of 80 percent. And when official data are adjusted, rural-urban disparities accounted for more than half of total inequality in 1995 and explain even more of the increase since 1985.

China's large rural-urban gap points to imperfect mobility in factor markets, especially for labor. Despite increasing accommodation of the swelling demand for rural emigration, important impediments remain, reflecting the government's desire to control the pace of migration and to ensure grain self-sufficiency. The absence of a housing market and limited access to social services in urban areas pose additional constraints to labor mobility.

Regional disparities are widening

As China opened to the outside world, the coastal provinces were poised to seize opportunities presented by their proximity to world markets, access to better infrastructure, and educated labor force. But they were also helped by the central government's preferential policies, which stimulated foreign investment. As a result interprovincial inequality has risen. It accounted for almost a quarter of total inequality in 1995 and explained a third of the increase since 1985 (see figure 1). In 1985 residents of interior China earned 75 percent as much as their coastal counterparts; by 1995 this had dropped to 50 percent.

Access to opportunities is becoming less equal

People's different endowments suggest that inequality in outcomes is not only unavoidable but also that it can help nourish creativity and spur growth. As a result most societies tolerate some inequality in income. How much depends on the historical and cultural factors shaping each society's preferences. Much of the increase in China's income inequality needs to be evaluated in the context of the country's systemic transition. Transition has brought an adjustment in relative prices, revaluing endowments and characteristics that are conducive to productivity gains. Such adjustments are acceptable.

More insidious is inequality in access to opportunities to improve incomes and welfare, which also has been found to hamper growth prospects. China's highly egalitarian land distribution has helped protect the nutritional status of the poor. But educational attainment and access to health care are becoming less equal as market orientation encourages cost recovery in public institutions. There is also evidence that families invest less in girls' than boys' education and health. Coupled with rising discrimination against women in the labor market, this tendency threatens to erode women's hard-earned gains, which have been a source of national pride. Finally, imperfect labor mobility creates unequal access to better-paying jobs. China's segmented labor markets are reflected in the near-absence of urban poverty, the relatively low level of urbanization, and the large rural-urban income gap.

Policies to grow with

Income inequality may well continue to rise as China's transition unfolds. But increasing inequality need not undermine growth or social harmony—so long as growth is broadly based, policy biases are eliminated, and the poor and vulnerable are protected.

Protecting the poor and the vulnerable

Investment in human capital is key to long-term improvements in welfare for all, but other policies can usefully differentiate treatment by segments of the population.

The absolute poor. In 1995 there were 70 million absolute poor in China. If current assistance programs were targeted more accurately, they would alleviate more poverty and cost less. In 1990 almost half of China's poor lived outside the counties designated for special assistance programs. These programs would be more effective if they were targeted at the level of townships, or perhaps even administrative villages

The government should also consider refocusing priorities in its poverty reduction strategy. A renewed emphasis on basic education and health services for the poor is essential, combined with assistance for finding employment in economically advanced areas. There is a need to ensure essential health services for the poor and to strengthen public health programs. Poor households must be compensated (through scholarships) for the costs of educating their children, and in this the government is aided by the demographic transition—the number of school-age children is declining. Government assistance to the poor in finding jobs outside their immediate home area should be expanded because remittances contribute significantly to reducing rural poverty.

The near poor. About 100 million additional people survive on less than $1 of income a day (in 1985 purchasing power parity dollars) and derive almost half their incomes from grain, a heavily regulated subsector. Reforms in grain policies are needed to improve this group's standard of living. Greater integration in labor markets and better-functioning credit markets would also help. The government's decision to align grain procurement prices to market prices is welcome. Better transport infrastructure and changes in the grain distribution system would help boost farmgate prices, and more spending on agricultural research and extension could increase yields. Above all, the near poor would benefit from shifting out of low-return grain production into higher-value crops or off-farm employment. But such shifts would require government willingness to import more food.

Urban poverty. Although urban poverty is negligible, it may become an increasing concern as enterprise reforms deepen and China continues to urbanize. Unemployment (including furloughs) in China's cities has already reached 8 percent of the labor force. The government needs better information about the urban poor to develop assistance programs for them. Establishing a meaningful urban poverty line would help, as would systematic monitoring of the unemployed. Now is also an opportune time for the government to examine its social protection system; substantial work has already gone into analyzing the pension and health care finance systems. Additional efforts should concentrate on other benefits such as unemployment compensation, disability, and labor training and retraining schemes. Finally, a better job information system would facilitate the redeployment of labor, while a systematic evaluation of urban job creation programs would help disseminate the lessons of their success or failure.

Eliminating policy biases and strengthening regulations

Public policies in China tend to exacerbate the gap between rich and poor. Policy changes in four areas would benefit welfare and income distribution.

Redressing the urban bias. Housing, food, migration, credit, state employment, and other policies provide de facto subsidies for urban residents. Some of these policies directly lower the welfare of rural residents. Others do so indirectly, by preempting public resources that could be targeted at more needy populations.

Removing the coastal bias in economic policies. The natural and human capital advantages of the coastal provinces are sufficient to attract foreign investment and need not be bolstered with preferential policies. In addition, a reformed intergovernmental transfer scheme would reduce disparities in public spending across provinces; the government should accelerate its design and implementation. Policies that favor the interior also may help address the widening gulf between China's interior and the coast, but additional research is needed on an appropriate package of regional growth policies. International experience with regional development efforts has generally been negative, but there has been little systematic analysis of this important issue.

Countering gender bias in household allocation decisions and in the marketplace. Education grants can provide incentives for families to educate girls. Government can make the retirement age for men and women the same and avoid discrimination in benefits provision. Regulations and firm-level subsidies can spread the costs of child rearing, which usually are shouldered solely by women.

Dealing fairly with the rich. Some of China's newly rich have worked hard and taken calculated risks to benefit from new market opportunities. But others are taking advantage of China's incomplete transition to accumulate ill-begotten wealth. The government is right to focus on the second group. To combat corruption and to counter rent-seeking behavior, the government must enforce its regulations. Doing so will require reducing bureaucratic discretion, establishing clear and transparent rules for public decisionmaking (such as public procurement), and stamping out access to insider information in financial markets.

chapter one

Richer but Less Equal

As China's reforms have continued, its income distribution has become more unequal. In 1981 China was an egalitarian society, with an income distribution similar to that of Finland, the Netherlands, Poland, and Romania. But rapid economic growth has brought dramatic change, so that China's income inequality is now just about average by international standards (figure 1.1). In 1981 China's Gini coefficient (a measure of inequality of income distribution ranging from 0, absolute equality, to 100, absolute inequality) was 28.8. By 1995 it was 38.8—lower than in most Latin American, African, and East Asian countries and similar to that in the United States, but higher than in most transition economies in Eastern Europe and many high-income countries in Western Europe.

The increase in China's Gini coefficient was by far the largest of all countries for which comparable data are available (figure 1.2). Such a large change is unusual. Levels of inequality vary enormously by country, but income distributions are strikingly stable over time within a given country (Deininger and Squire 1996). When large changes do occur, they generally signal deep structural transformations in the underlying distribution of assets and in their rates of return. Recent examples come mainly from transition economies, but Brazil, Thailand, and the United Kingdom have also experienced substantial increases in inequality.

FIGURE 1.1

Since 1981 China's income distribution has become much less equal . . .

Note: See figure notes at end of chapter.
Source: Deininger and Squire 1996; World Bank 1996f; World Bank staff estimates.

FIGURE 1.2

. . . because of remarkable changes between 1981 and 1995

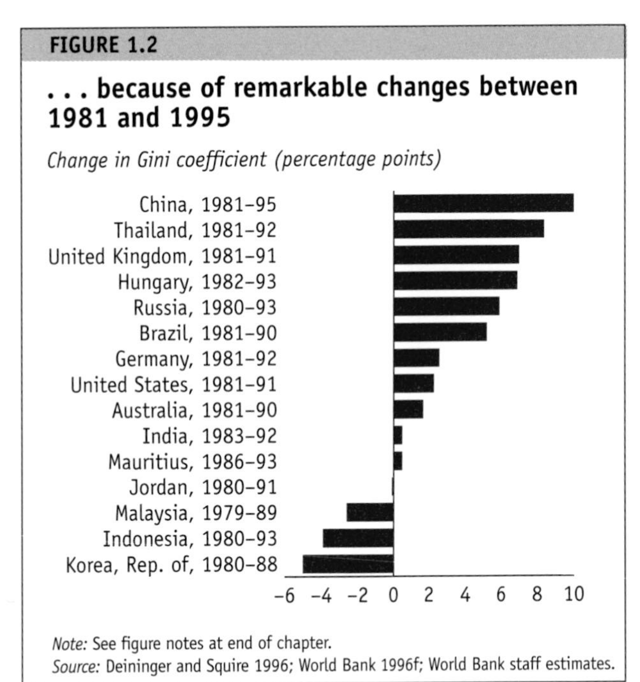

Note: See figure notes at end of chapter.
Source: Deininger and Squire 1996; World Bank 1996f; World Bank staff estimates.

Still, China's recent experience stands out even in this crowd. Not even the transition economies of Eastern Europe and the former Soviet Union registered increases in inequality as large as those observed in China over the past fifteen years. Moreover, some East Asian countries actually saw inequality fall during this period.

Growing unequal: National trends

Regardless of how inequality is measured, China's income distribution has become more unequal (figure 1.3). This conclusion holds despite the many shortcomings of China's household survey data (box 1.1). The decile ratio (the ratio of the mean income of the top 10 percent of the population to the mean income of the bottom 10 percent) has been rising, especially since 1990, suggesting increasing divergence between the richest and poorest groups.

Since the start of reforms China has experienced three distinct periods in the evolution of personal incomes (box 1.2).[1] Between 1981 and 1984 all segments of society benefited from across-the-board improvements in welfare, with only a small rise in inequality. Between 1984 and 1989 personal incomes stagnated and became increasingly unequal, implying real losses in the standard of liv-

FIGURE 1.3

Other measures of inequality point to the same conclusion as the Gini index, 1981–95

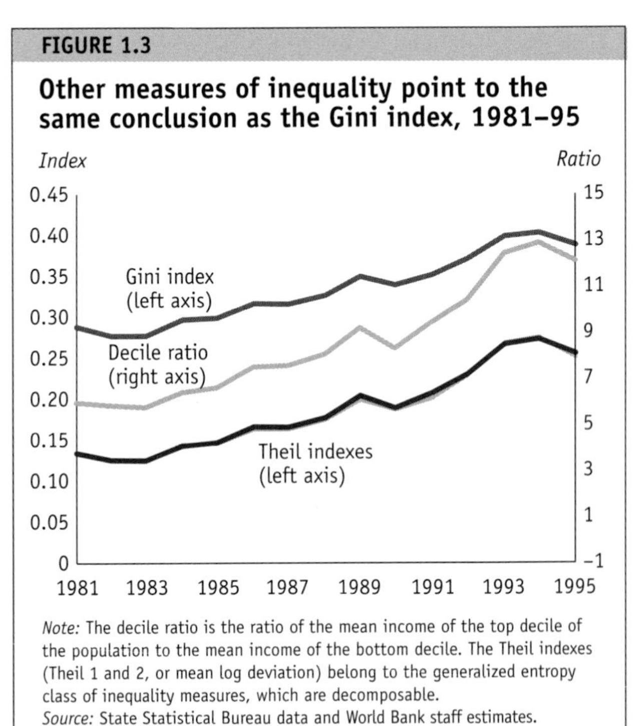

Note: The decile ratio is the ratio of the mean income of the top decile of the population to the mean income of the bottom decile. The Theil indexes (Theil 1 and 2, or mean log deviation) belong to the generalized entropy class of inequality measures, which are decomposable.
Source: State Statistical Bureau data and World Bank staff estimates.

BOX 1.1

Shortcomings of household survey data—and what this report does about them

This report's analysis of China's income distribution is based on the results of household surveys carried out by China's State Statistical Bureau. These surveys have many shortcomings, but they remain the only source of comprehensive data over a long enough period to assess national trends in the distribution of income (see World Bank 1992 and Chen and Ravallion 1996 for a detailed discussion of problems with rural surveys). The main concerns with the surveys relate to coverage, definitions, and processing after data collection:

- The surveys are based on the registration system (*hukou*) and so do not capture migrants into urban areas without a hukou. Few migrants acquire resident status, so this omission is serious and growing.
- Urban and rural surveys are based on incompatible definitions of incomes, which reduces comparability and hinders aggregation into a national distribution.
- The data do not account for spatial differences in the cost of living. Thus neither regional differences within the urban and rural surveys nor national rural-urban differences can be treated systematically.
- Urban household surveys exclude in-kind income such as housing, health care, and education benefits. Also, the surveys appear to be geared toward recording labor income, and so miss many of the newly affluent.
- Summary urban data in the *China Statistical Yearbook* for 1989–95 suffer from aggregation problems that understate urban inequality.
- Until 1990 rural household surveys valued in-kind grain income at official prices, understating rural income considerably. After 1990 and until recently own-grain consumption was valued at the weighted average of official and market prices, but practice varied by province. Both distortions make it difficult to analyze trends over time and across provinces.
- Definitions of *residence* and *income* have changed over time. Urban residency was extended to some periurban areas in 1985, and pensioners were included in income surveys starting only in 1985.

This report did not have access to individual household data except for rural data from four Southern provinces (Guangxi, Guizhou, Guangdong, Yunnan) for 1985–90 and rural and urban data for Sichuan and Jiangsu provinces for 1990. As a result systematic corrections could not be made to the shortcomings identified above. Instead, partial adjustments have been made in various sections of the report to indicate the magnitude and direction of the resulting effect on inequality. The aggregate effect of these and other necessary corrections cannot be determined with any precision at this time. Further collaboration with the State Statistical Bureau is needed to confirm that the report's findings are robust and to adjust survey design and tabulation methodology for the future. The report's analysis includes the following adjustments:

- The living standards of migrants are discussed only with reference to special surveys on migrant populations and cannot be integrated with the overall income distribution.
- For the most part, national trends are based on an aggregation of rural and urban household surveys into a national distribution without any adjustments in the definition of income or for spatial price differentials.
- Some indicative adjustments help provide a more accurate picture of the components of inequality: mean urban and rural incomes are adjusted to include in-kind incomes using information from the State Statistical Bureau and the four-province rural dataset; and a cost of living differential is introduced to account for higher prices in urban areas.
- The four-province dataset is used to correct for grain pricing, cost of living differentials, and the valuation of housing and consumer durables. This allows for a more accurate valuation of rural incomes, inequality, and changes over time.
- For 1989–95 the report uses urban data aggregated by the Beijing office of the State Statistical Bureau but coming from a subsample of the survey that has consistently higher mean incomes than the published data.
- The effect of in-kind incomes on levels of and changes in urban inequality is investigated using information provided by the State Statistical Bureau for 1990 and 1995.
- The analysis of the determinants of inequality relies largely on available microdata and thus is sensitive to measurement changes.

ing for a large part of the population. Renewed growth in incomes between 1990 and 1995 appears to have reached the poorer (but not the poorest) segments of society but was accompanied by substantial increases in inequality.

Growing out of poverty?

China's record on reducing poverty is enviable. Since reforms started in 1978, China has lifted some 200 million people out of absolute poverty. Most of this progress occurred in the early years of reforms, when the introduction of the household responsibility system transformed China's countryside. In the mid-1980s and early 1990s poverty levels stagnated (and increased in some years) despite steady gains in per capita GDP. These trends generated concern about the quality of Chinese growth because increases in inequality that occur because the poor stay poor or get poorer while

BOX 1.2

China's income distribution, 1981–95: A tale of three periods

Growth with equity

China started its economic reforms in 1978 with the introduction of the household responsibility system. The unleashing of rural productivity in response to the provision of incentives for personal gain are by now well-known. Between 1981 and 1984 the national income distribution shifted to the right (see top figure), indicating across-the-board benefits from reforms. Mean incomes increased by 12.6 percent a year (in real terms) during this period. The slightly flatter curve in 1984 indicates an increase in inequality relative to 1981, although the distribution of income remained remarkably equal for such a large shift in average incomes. Between 1981 and 1984 the Gini coefficient increased slightly, from 28.8 to 29.7

Inequality with little growth

Between 1984 and 1989 the income distribution curve shifted dramatically. Inequality became much more pronounced, with the shorter and wider 1989 curve reflecting a jump in the Gini coefficient from 29.7 in 1984 to 34.9 in 1989 (see middle figure). Interestingly, these large distributional shifts occurred despite stagnation in personal incomes. Between 1984 and 1989 average incomes increased by less than 1 percent a year. Although the mean income of the top decile of the population increased by 2.8 percent a year (shift to the right in the right tail), the mean income of the bottom decile dropped by 4.5 percent a year (shift to the left in the left tail). Positive income growth started occurring only with the sixth decile. These changes are reflected in a deterioration in poverty indicators during this period, which also saw an increase in rural-urban disparities (the bulge in the upper right side of the 1989 curve).

Growth with inequality

Between 1990 and 1995 renewed growth in personal incomes (7.1 percent a year) was associated with substantial increases in inequality (see bottom figure). During this period the Gini coefficient increased from 33.9 to 38.8. Still, the benefits of growth reached people at the lower end of the income distribution, with the possible exception of those at the very bottom (incomes less than 190 yuan a year in 1990 prices). Incomes of the bottom decile (less than 337 yuan a year) increased by 1.7 percent a year between 1990 and 1995, but most of these gains were registered in 1994 and 1995, when the mean income of this group grew by 6.7 percent a year. Mean incomes of the top decile increased by 9.7 percent a year between 1990 and 1995 but by as much as 12.1 percent between 1990 and 1994. Growth in 1995 appears to have considerably equalized the distribution of incomes.

Source: State Statistical Bureau data and World Bank staff estimates.

Between 1981 and 1984 Chinese from all walks of life benefited from reform

Share of population (percent)

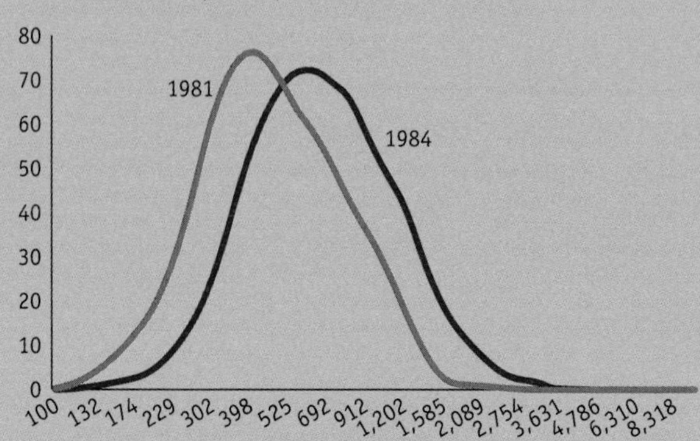

Income (1990 yuan, log scale)

Source: World Bank staff estimates.

Between 1984 and 1989 the rich got richer as the poor got poorer

Share of population (percent)

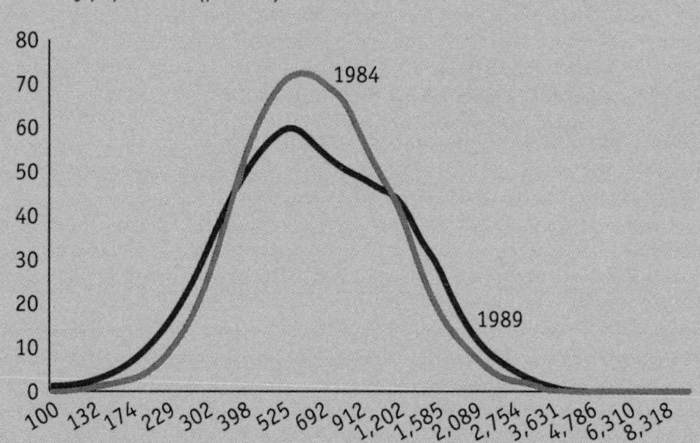

Income (1990 yuan, log scale)

Source: World Bank staff estimates.

Between 1990 and 1995 most people got richer but inequality grew substantially

Share of population (percent)

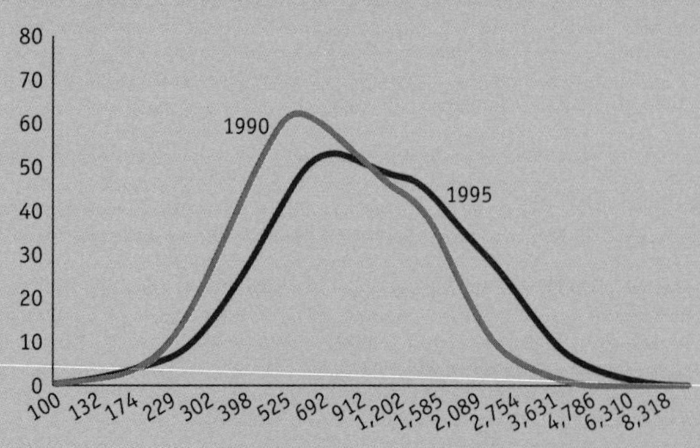

Income (1990 yuan, log scale)

Source: World Bank staff estimates.

the rich get richer are particularly damaging. Since 1992, however, poverty has declined markedly, and at the end of 1995 less than 6 percent of the population had incomes below the absolute poverty line.

Overall, GDP growth helped reduce poverty during this period (figures 1.4 and 1.5). Between 1981 and 1995 the poverty elasticity of growth was close to –0.5, implying that the number of poor declined by 0.5 percent for each percentage point of growth in GDP per capita. But this outcome masks substantial differences in the effect GDP growth had on poverty during different periods. During 1981–84 and 1993–95 per capita GDP growth was high (about 10 percent a year) and poverty elasticities were –3.6 and –1.7, respectively. Between 1985 and 1992 per capita GDP growth was lower but still impressive (7.4 percent a year) while the poverty elasticity was (slightly) positive, implying that the number of poor people increased during this period.

FIGURE 1.4

Poverty has declined dramatically since the start of reforms . . .

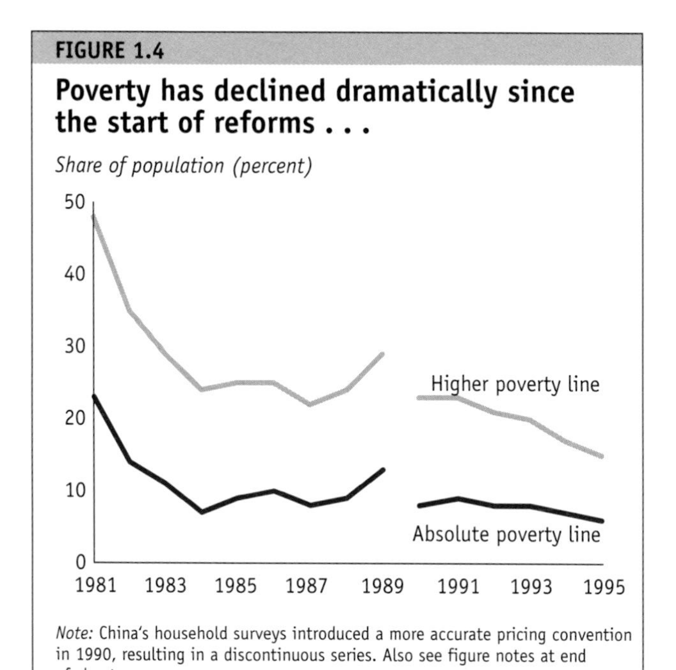

Note: China's household surveys introduced a more accurate pricing convention in 1990, resulting in a discontinuous series. Also see figure notes at end of chapter.
Source: State Statistical Bureau data and World Bank staff estimates.

FIGURE 1.5

. . . but the relationship between per capita GDP growth and poverty has been unstable

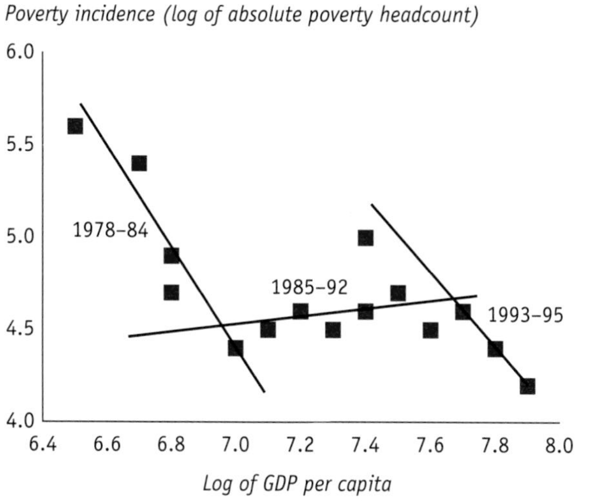

Poverty incidence (log of absolute poverty headcount)

1978–84

1985–92

1993–95

Log of GDP per capita

Note: See figure notes at end of chapter.
Source: State Statistical Bureau data and World Bank staff estimates.

Poverty in China is a rural phenomenon. Even in 1981 just 0.3 percent of the urban population lived in absolute poverty, while 28.0 percent of the rural population did. Thus rural growth is likely to be more important to reducing poverty than aggregate growth, especially since rural-urban migration is limited. In fact, during 1991–95 the poverty elasticity of rural per capita income growth was high and did not display the variation observed for per capita GDP growth. But when rural income growth stagnated in 1985–92, poverty alleviation stalled (figure 1.6). When GDP growth translated into growth in *rural* per capita incomes before 1985 and after 1992, however, the poor benefited substantially (figure 1.7), suggesting that rural income growth was distributed relatively evenly.

One of the most curious aspects of China's development during 1985–92—and one that requires further investigation—is the divergence between per capita GDP growth and personal income growth. Stagnation in personal incomes during this period affected both rural and urban populations, such that the share of personal incomes in GDP fell from a peak of 60 percent in 1984 to 45 percent in 1993.[2] There are several possible explanations. If these data accurately depict trends, there should have been large increases in enterprise and government savings during this period; available data do not support this view, however. More likely, personal income growth is being understated (in part because migrants' expeditures in urban areas are being left out) or GDP growth is being overestimated (probably due to an underestimation of deflators).

Rural income growth has been vital to reducing poverty. Disaggregating the effects of growth and redistribution on poverty shows that the number of poor in China would have increased by 50 percent in the absence of rural growth because of adverse distribu-

FIGURE 1.6

Per capita GDP growth did not always yield increases in personal incomes . . .

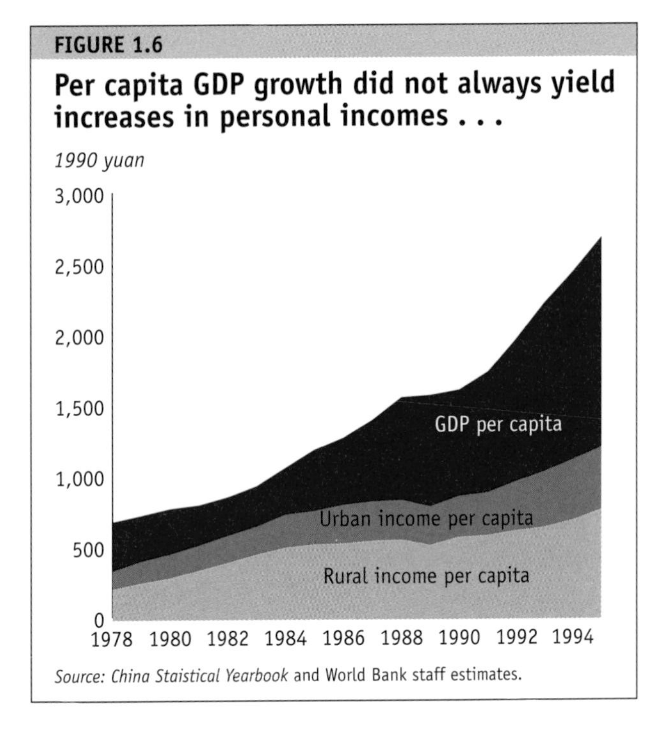

Source: China Staistical Yearbook and World Bank staff estimates.

TABLE 1.1

Income redistribution simulations: Results for poverty and inequality, 1990

Indicator	Before	After
Gini coefficient	33.9	30.7
Rural	29.6	29.6
Urban	22.4	22.4
National mean income (yuan)	888	888
Rural	684	753
Urban	1,457	1,266
Poverty incidence (percentage of population)	8.3	6.0
Rural	11.3	8.1
Urban	0.0	0.1

Note: Calculations simulate a 10 percent increase in rural incomes through a 15 percent tax on urban incomes and assume no transfer losses.
Source: State Statistical Bureau data and World Bank staff estimates.

FIGURE 1.7

. . . but when it boosted rural incomes, poverty declined

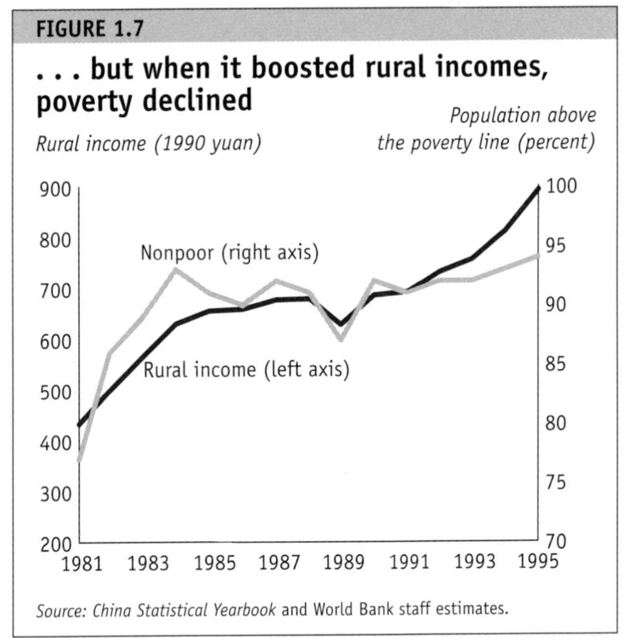

Source: China Statistical Yearbook and World Bank staff estimates.

FIGURE 1.8

The importance of growth, growth, and more growth

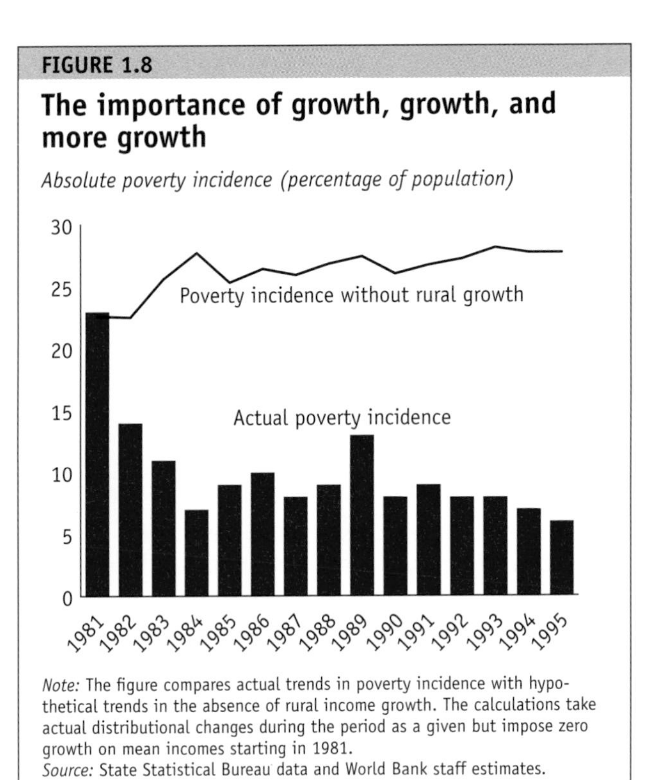

Note: The figure compares actual trends in poverty incidence with hypothetical trends in the absence of rural income growth. The calculations take actual distributional changes during the period as a given but impose zero growth on mean incomes starting in 1981.
Source: State Statistical Bureau data and World Bank staff estimates.

tional changes (figure 1.8) Since the start of reforms there have been only two years when distributional shifts appear to have favored the poor—1985 and 1990—and these are likely to be measurement effects because there were large adjustments to the State Statistical Bureau's household surveys in both years.[3]

These experiences suggest that redistributive policies should remain secondary in China's poverty reduction strategy. Given the political difficulties associated with effecting large distributional shifts and the possible negative effect such shifts would have on growth, China should raise rural incomes through growth rather than through redistribution. In this regard measures to augment poor people's assets (land and human capital) are essential to reduce poverty and to achieve a more equal distribution of income. Still, well-targeted programs will continue to be needed to reach those who may be bypassed by the forces of growth. Thus policies should continue to improve the health and education of the poor, facilitate access to markets, and enhance labor mobility.

A simple calculation showing the effects of income redistribution from urban to rural areas demonstrates the limits of redistributional policies (table 1.1). If 15 percent of 1990 urban incomes had been redistributed to rural residents, the incidence of poverty would have fallen from 8.3 percent of the population to 6.0 percent and the Gini coefficient would have dropped from 33.9 to 30.7. But the same results could have been achieved in just two years if the incomes of the bottom decile of the rural population had grown by 5 percent a year. What actually happened between 1993 and 1995, in fact, closely mirrored the results of this simulation: broadly based growth in rural incomes during those two years lowered the incidence of poverty from 8.2 to 5.7 percent of the population, with modest declines in inequality.

Notes

1. The density distributions used to analyze these periods were generated by software designed to process distribution data—the Gini ToolPak—developed by Yuri Dikhanov of the World Bank. State Statistical Bureau data in the *China Statistical Yearbook* include tabulations for the share of households with per capita income within a range, for rural households; and average per capita income for each decile (each 5 percent starting in 1989) of households, with corresponding share of total income ranked by per capita income, for urban households. The computations here convert household distributions into population distributions based on household size per income category, shown in the *China Statistical Yearbook* for the urban survey and supplied by the State Statistical Bureau for the rural survey (for 1985, 1990, and 1992–95, with interpolations for intervening years). Incomes are deflated using the urban or rural consumer price index (except for pre-1985 rural data, which were deflated by the rural retail price index). Rural and urban data are aggregated into the national distribution using population data (based on registration status) from the *China Statistical Yearbook*.

2. Personal incomes are calculated using household survey data. The State Statistical Bureau's calculation of personal incomes from national accounts data shows the same ratio rising from 61 percent in 1984 to 68 percent in 1993.

3. In 1985 the rural sample survey was doubled in size and the concept of income became more comprehensive. In 1990 changes were made to the valuation of own-grain consumption.

Figure notes

Figures 1.1. and 1.2 The countries in the figure are chosen largely on the basis of availability of comparable data, but also with a view to representing different regions. Comparable statistics on income distribution are still not common and restricted the number of countries that could be included. All Ginis shown are based on income (not expenditure) distributions. China data are based on World Bank staff estimates. Data for all other countries are for years between 1988 and 1992.

Figures 1.4 and 1.5 The incidence of poverty is calculated by applying the poverty line to a constant price (1990) distribution of income per capita. The rural and urban consumer price indexes are used to convert current income into constant 1990 prices. Because the rural consumer price index is available starting only in 1985, the rural retail price index is used for previous years. The absolute poverty line, established at 318 yuan in 1990 prices, reflects the income required to meet minimum nutritional (2,100 calories a day) and nonfood requirements (see World Bank 1992) and corresponds to about $0.70 a day in 1985 purchasing power parity (PPP) dollars, using data available in the Penn World Tables (see Summers and Heston 1991). The higher poverty line is set at 454 yuan in 1990 prices, equivalent to $1 a day in 1985 PPP dollars. Given this report's focus on income inequality, the $1 a day standard was applied to the income distribution even though the international standard developed by the World Bank to monitor progress in poverty reduction applies the poverty line of $1 a day (in 1985 PPP dollars) to consumption expenditure. In the absence of a consumption distribution for China, the latter methodology involves shifting the income distribution by the average ratio of consumption to income (see World Bank 1996e and Ahuja and others 1997). Doing so yields accurate results so long as the slope of the Lorenz curves for consumption and income at the poverty line are the same, which appears to be the case for 1992 (a year for which consumption expenditure data were available). In 1995 there were 170 million people with *incomes* below $1 a day (in 1985 PPP dollars) but as many as 270 million who *consumed* less than $1 a day (in 1985 PPP dollars).

chapter two

Growing Apart: Rural-Urban and Coastal-Interior Gaps

A look at the components of the worsening national income inequality reveals unique features in China's income distribution and points to the unfinished nature of transition. The widening gulf between rural and urban incomes is the biggest contributor to increased inequality. Regional disparities are responsible for a smaller but growing portion of inequality.

The rural-urban divide is growing

The income gap between China's rural and urban populations is large and growing. According to State Statistical Bureau data, rural-urban disparities accounted for more than one-third of inequality in 1995 and about 60 percent of the increase in inequality between 1984 and 1995 (figure 2.1). Adjusting these data for some of the shortcomings noted in box 1.1 reveals an even starker picture. Adjusted, rural-urban disparities accounted for more than 50 percent of inequality in 1995 and explain 75 percent of

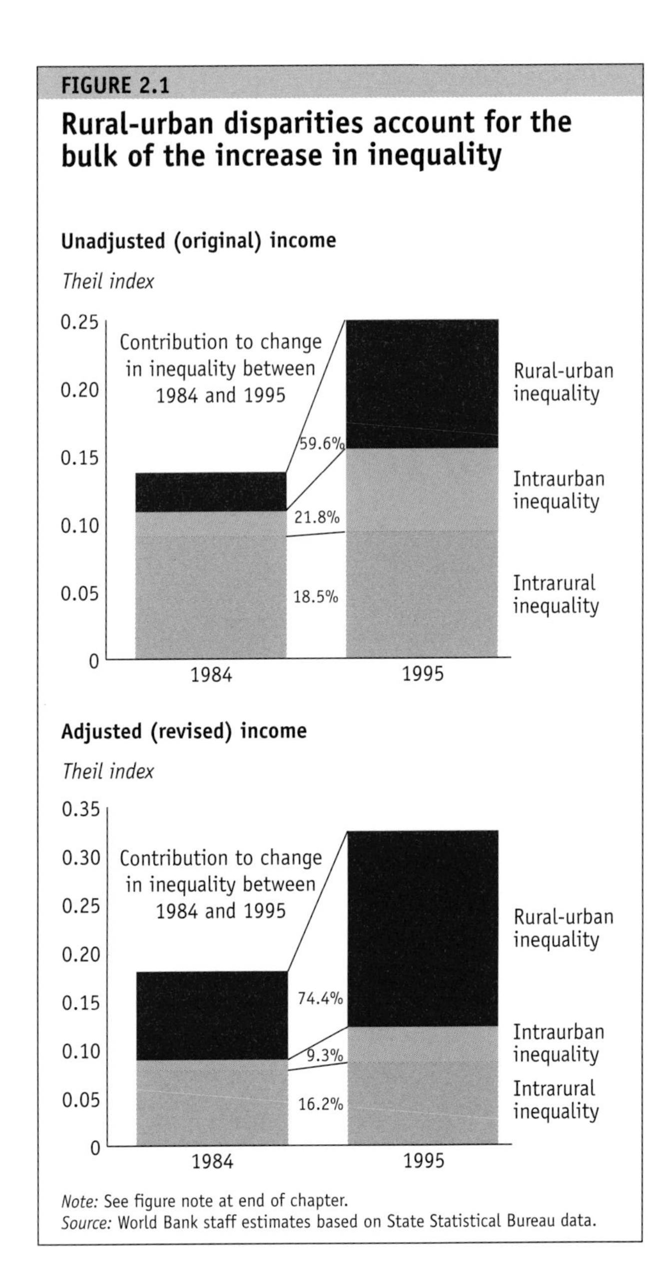

Note: See figure note at end of chapter.
Source: World Bank staff estimates based on State Statistical Bureau data.

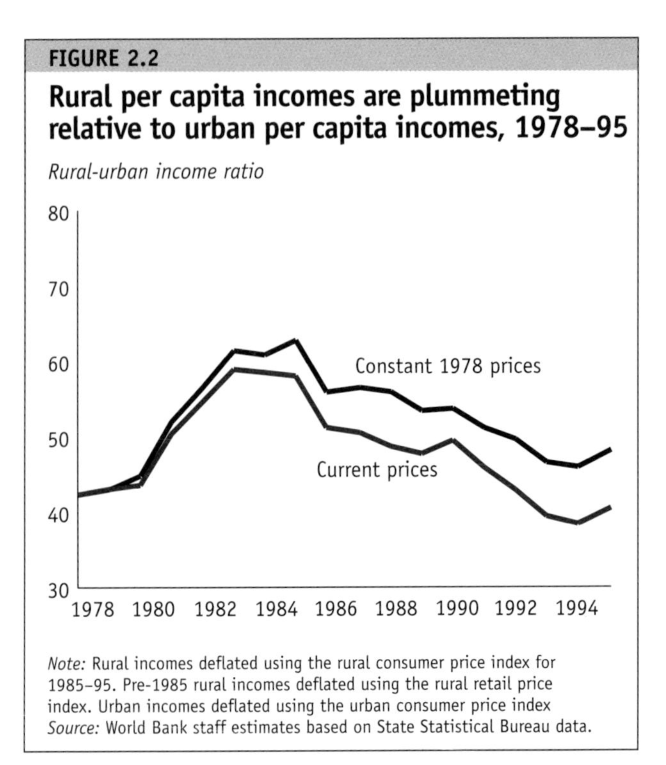

Note: Rural incomes deflated using the rural consumer price index for 1985–95. Pre-1985 rural incomes deflated using the rural retail price index. Urban incomes deflated using the urban consumer price index
Source: World Bank staff estimates based on State Statistical Bureau data.

the increase between 1984 and 1995. The data adjustments yield two important changes: they lower inequality within rural and within urban areas but maintain the trend increase, and they magnify rural-urban disparities. The overall impact of the changes is an increase in total inequality.

China's rural-urban income gap is large by international standards. Data for thirty-six countries show that urban incomes rarely are more than twice rural incomes; in most countries rural incomes are 66 percent or more of urban incomes (Yang and Zhou 1996). In China rural incomes were only 40 percent of urban incomes in 1995, down from a peak of 59 percent in 1983 (figure 2.2). These figures do not take into account the differential increases in the cost of living between urban and rural areas. But even the deflated series reveals an unmistakable trend. Rural incomes grew rapidly during the early years of reform but in 1985 began to trail increases in urban incomes. This trend reversed only in 1995.

Two other variables affect the accurate assessment of rural-urban income disparities, and both have been incorporated in the adjusted data in figure 2.1 and table 2.1: cost of living differences between rural and urban areas and the underestimation of both rural and urban in-kind income.[1] Rural incomes were adjusted to include imputed rent and urban incomes to include in-kind income for housing, education, health care, pensions and other subsidized services. In addition, a 15 percent cost differential was introduced between urban and rural areas. These adjustments lowered rural incomes to 31 percent of urban incomes in 1990—substantially less than the 50 percent suggested by official data.[2] The adjusted data also yield much higher national inequality (as measured by the Gini coefficient) because urban income increases more than compensate for the higher cost of living in urban areas.

The magnitude of the gap between China's rural and urban incomes points to imperfect mobility in factor markets, especially for labor. Despite increasing accommodation of the swelling demand for rural emigration, impediments to labor mobility remain. These are motivated by the government's desire to control the pace of migration and ensure grain self-sufficiency. The costs of relocation, lack of job information, absence of a housing market, and limited access to social services in urban areas pose additional constraints to migration. Meanwhile, government policies continue to prop up urban standards of living. Urban citizens are subsidized in a variety of ways, including through the absence of hard budget constraints for state-owned enterprises (primarily to protect urban jobs), low-cost capital for urban enterprises, low-cost housing for urban residents, and generous pensions and health insurance schemes. Enterprise and financial sector reforms and fiscal constraints are challenging these acquired rights: some in-kind benefits have been eliminated while others are being monetized, as workers now pay higher rents and contribute more to their pension and medical benefits. This may account for part of the observed increase in the rural-urban income gap in official data.

The magnitude of rural and urban inequalities depends on how incomes are measured

According to official data, both rural and urban inequality increased steadily between 1981 and 1995 (figure 2.3). The urban Gini coefficient increased from 17.6 in 1981 to 27.5 in 1995, although it dropped during the recession years of 1989–91 and in 1995. The rural Gini increased from a much higher base of 24.2 in 1981 to 33.3 in 1995. It dipped in 1985 and in 1990 and has stabilized since 1993. The declines in 1985 and 1990 almost certainly represent measurement effects because there were large adjustments to the State Statistical Bureau's household survey in both years.

This section examines the effect data adjustments have on urban and rural inequality: corrections are made to coverage, valuation, and price differentials.[3] Revisions to the rural household survey data raise the mean income and reduce inequality but do not alter the finding that overall inequality increased between 1985 and 1990. Similarly, incorporating in-kind income yields substantially higher urban incomes that are more equally distributed, but urban inequality clearly rose between 1990 and 1995.

TABLE 2.1

Rural-urban income gap and inequality with data adjustments, 1990

Meassure	Rural-urban income ratio (percent)	National Gini coefficient	Contribution to national inequality (percent)
Official data	49.5	33.9	29.5
With 15 percent higher cost of living in urban areas	56.9	31.9	20.5
Plus in-kind incomes	30.5	40.6	51.8

Source: State Statistical Bureau urban household survey team and World Bank staff estimates.

FIGURE 2.3

Rural and urban inequality have increased steadily since reforms began

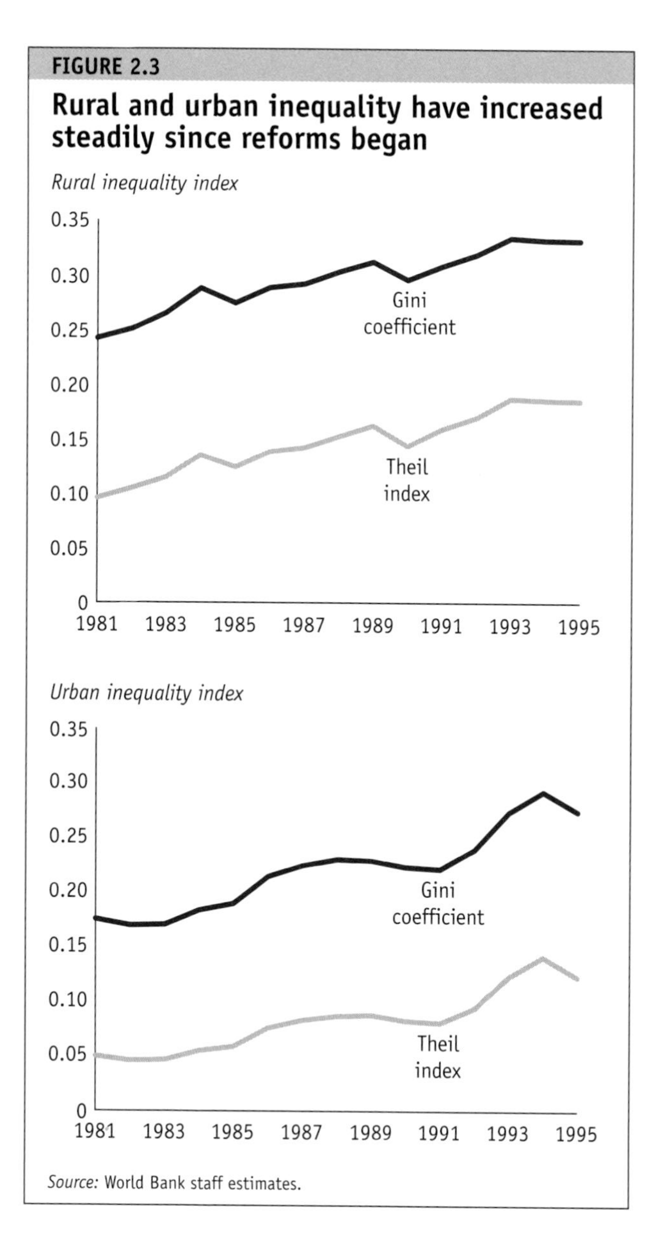

Source: World Bank staff estimates.

Rural incomes revalued

Official data from the rural household surveys prior to 1990 rely on administrative planning prices for the valuation of in-kind income from consumption of own-farm production. This approach undervalues a large component of income—nonmarketed home production of grain—and at a rising rate over time. According to standard definitions, 21 percent of rural incomes in 1985–90 in the four-province data set (Guangxi, Guizhou, Guangdong, and Yunnan) came from grain production, of which 80 percent was the imputed value of consumption from own production.[4] Another problem is that the incomes used in the State Statistical Bureau's tabulations do not include imputed rents for housing and consumer durables. Past work also has ignored spatial differences in the cost of living.

To correct for these shortcomings, in-kind grain income was revalued at median local (county-level) selling prices for grain, as determined from primary household data. The administrative prices conventionally used for valuation were 72 percent of the median selling price in 1985 but had fallen to 48 percent by 1990, resulting in serious undervaluation of grain incomes. Other adjustments were made to impute rents for housing and consumer durables based on the asset valuations available in the primary survey data. And new province-level spatial and intertemporal cost of living indexes were constructed (based on poverty lines) to measure the local cost of the same standard of living everywhere.[5]

To assess the effect of these data adjustments, inequality indicators were calculated for each of three income definitions (figure 2.4). The first (original income) is the State Statistical Bureau's net income measure direct from its data files. The second incorporates imputed rents and the revaluation of grain income from own production. The third uses the new cost of living deflator as well. Although inequality increased during 1985–90 for all three income definitions, the adjusted

FIGURE 2.4

Changes in inequality are much less pronounced for adjusted data, 1985–90

Gini coefficient

35
30
25
1985 1986 1987 1988 1989 1990
Unadjusted (original) income
New valuation methods
New valuation plus new cost of living index

Source: Ravallion and Chen 1997.

FIGURE 2.5

Lorenz curves for China's rural south converge once data have been adjusted, 1985 and 1990

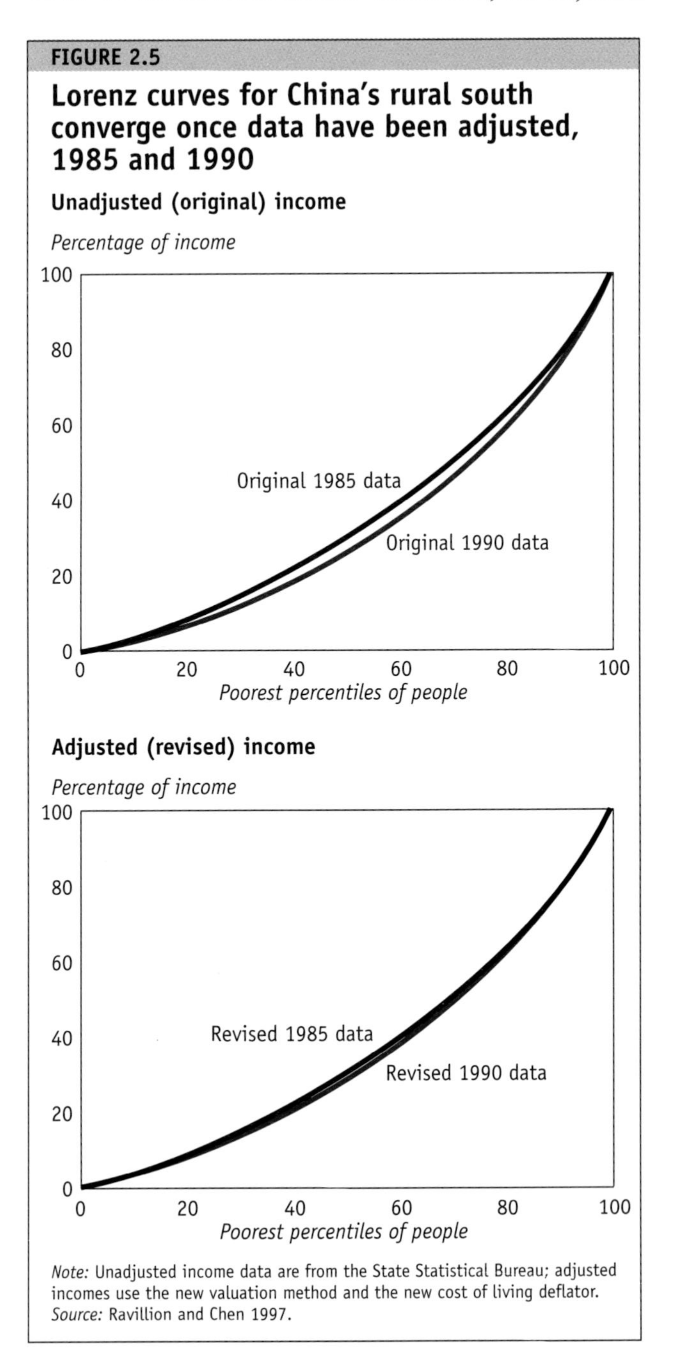

Note: Unadjusted income data are from the State Statistical Bureau; adjusted incomes use the new valuation method and the new cost of living deflator.
Source: Ravillion and Chen 1997.

data yield lower inequality and a lower rate of increase in inequality. Both conclusions are robust to the choice of inequality measure; with the revisions to the primary income data, the Lorenz curves for 1985 and 1990 converge (figure 2.5).[6] The revaluation of grain income in-kind accounts for most of the change, although the other changes also reduce inequality. The revaluation rates tend to be higher in 1990 than in 1985, largely reflecting the increasing divergence of market and planning prices.

Urban incomes revisited

The two most important shortcomings of the urban household surveys are coverage, which is restricted to *registered* urban residents, and undervaluation of in-kind income.[7] It is not possible to adjust for the first, increasingly important shortcoming, but crude estimates by the State Statistical Bureau help gauge the effect a more inclusive definition of urban income has on urban inequality. The largest adjustment is made for housing. Imputed rent for housing is calculated based on prices that approximate the market.[8] The income adjustment reflects the difference between this imputed rent and average rent per capita actually paid. To this amount is added individual contributions to the Provident Housing Fund deducted by employers (and thus not reflected in workers' income). The next largest adjustments are made for pensions and medical care. Because they are deducted by employers, pension contributions are not incorporated in the personal incomes reported in the survey.[9] The medical subsidy is calculated on the basis of an average 10 percent contribution to health care costs. Other benefits that accrue to urban residents but are not captured by the survey include price, education, and other subsidies (table 2.2).

Urban standards of living are much higher than official per capita income data suggest, thanks to declining but still large subsidies. Including the value of these in-kind benefits raises urban incomes 78 percent in 1990 and 72 percent in 1995 (table 2.3). Housing accounts for most (about 60 percent) of the increase. An important shortcoming of the urban household survey that has not been corrected for in this calculation is the omission of migrants (unregistered residents). Including them would reduce average urban incomes and thereby lower the rural-urban income gap.

TABLE 2.2

Total urban income, 1995

(yuan per capita)

		Percentiles of households, ranked by per capita income					
Type	Total	Bottom 10%	10–30%	30–50%	50–70%	70–90%	Top 10%
Income used for expenses[a]	4,612	1,777	2,733	3,592	4,572	6,153	10,250
In-kind income	3,304	2,076	2,803	3,284	3,629	4,030	3,882
Housing subsidy	1,960	1,182	1,705	2,047	2,267	2,353	1,906
Pension subsidy	595	233	380	495	603	853	1,222
Medical subsidy	306	226	264	295	325	366	367
Education subsidy	252	289	269	255	238	255	185
Communication subsidy	14	14	14	14	14	14	14
Price subsidy	59	59	59	59	59	59	59
Other in-kind income	87	69	83	88	91	95	95
Other welfare subsidy	31	24	29	31	32	35	34

a. State Statistical Bureau urban household survey definition.
Source: State Statistical Bureau urban household survey team.

TABLE 2.3

Distribution of in-kind income in urban areas, 1990 and 1995

(percentage of household survey definition of income)

		Percentiles of households, ranked by per capita income					
Year	Total	Bottom 10%	10–30%	30–50%	50–70%	70–90%	Top 10%
1990	77.7	137.5	106.3	90.5	79.4	67.1	49.8
1995	71.6	116.8	102.6	91.4	79.4	65.5	37.9

Source: State Statistical Bureau urban household survey team.

FIGURE 2.6

Adjusting urban incomes for in-kind benefits lowers urban inequality but does not alter the trend, 1990 and 1995

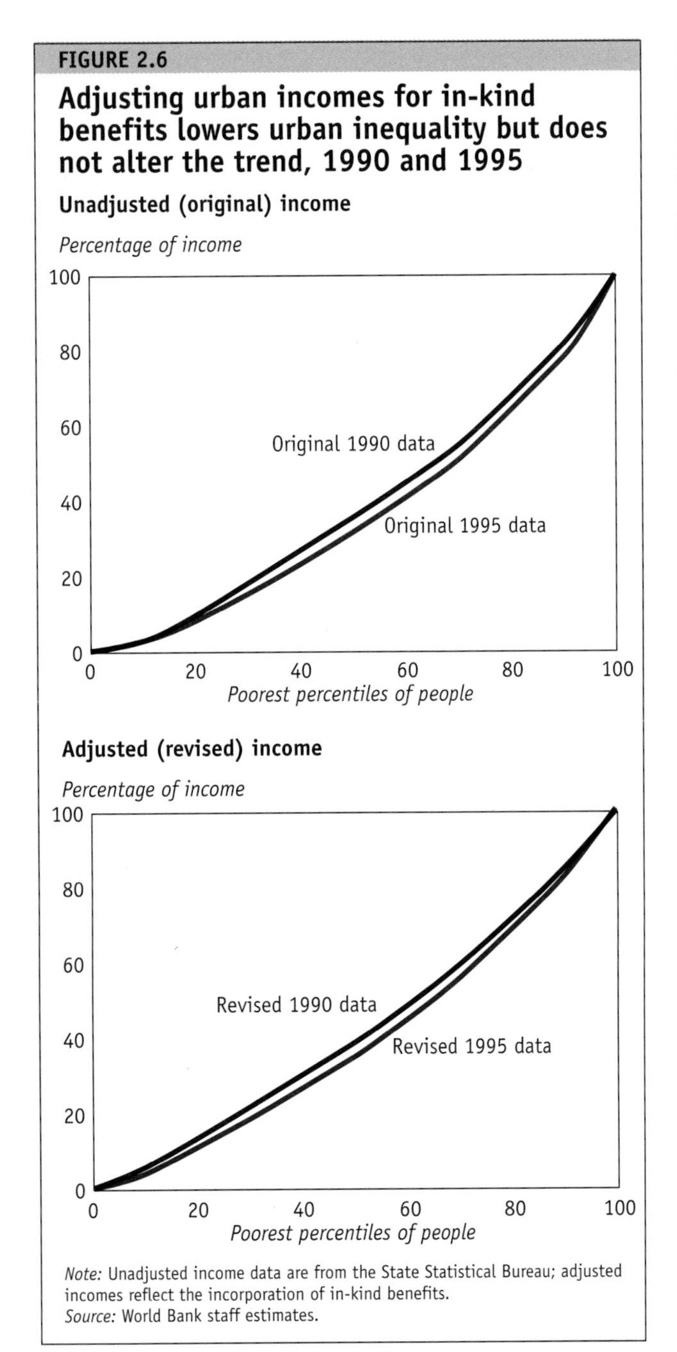

Note: Unadjusted income data are from the State Statistical Bureau; adjusted incomes reflect the incorporation of in-kind benefits.
Source: World Bank staff estimates.

FIGURE 2.7

Interprovincial and coastal-interior disparities are widening but remain moderate

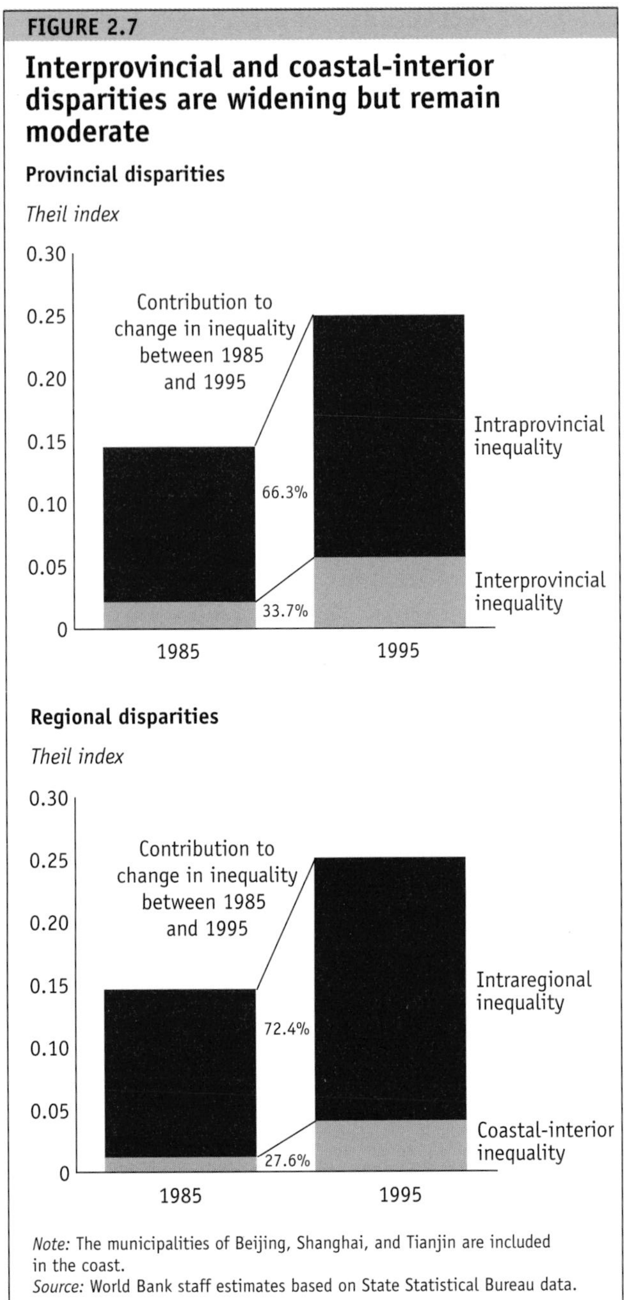

Note: The municipalities of Beijing, Shanghai, and Tianjin are included in the coast.
Source: World Bank staff estimates based on State Statistical Bureau data.

Whichever income measure is used, urban inequality increased between 1990 and 1995 (figure 2.6). But the distribution of in-kind benefits has an equalizing effect on urban welfare. The ratio of the top to bottom incomes (State Statistical Bureau definition) was 4.4 in 1990 and 5.8 in 1995, but once in-kind incomes are taken into account the ratios drop to 2.8 for 1990 and 3.7 for 1995. These are only approximations, however, as the largest adjustment (for housing) is based on a flat price per square meter and not market value, which would reflect the location and quality of housing.

The gulf between the coast and the interior is widening

Within China, much of the debate on inequality has focused on regional growth patterns. This is understandable given sharply widening regional disparities: interprovincial differences contributed 50 percent more to inequality in 1995 than in 1985, and the contribution of the coastal-interior gap doubled during the same period (figure 2.7 and box 2.1). Yet regional income inequality in China is still moderate. As much

BOX 2.1

Are China's provinces converging or diverging?

During 1981–94 per capita GDP growth in China's provinces averaged 8.9 percent a year. Excluding Qinghai, for which data are incomplete, Zhejiang Province grew the fastest at 13.1 percent a year; Heilongjiang grew the slowest at 5.1 percent a year—a virtual stall in China but respectable performance by any other standard. But variation in growth rates does not necessarily generate income disparities. If poorer provinces grow faster thanks to a "catching up" effect, incomes across regions would converge.

For all provinces and municipalities, per capita GDP converges (sigma convergence as measured by log deviation) until 1991 and diverges thereafter (see figure). The watershed year is 1992, when Deng Xiaoping made his famous trip to southern China, unleashing a period of rapid growth, especially in coastal areas. These results are qualitatively similar to those obtained by Jian, Sachs, and Warner (1996), but they show a much sharper decline in the standard deviation of log real per capita income between 1978 and 1991. This appears to be due to the use of different deflators.

But the results are sensitive to the treatment of Beijing, Shanghai, and Tianjin—the three municipalities that experienced lower than average per capita GDP growth during the period. If they are incorporated into their "natural" provinces, a divergent trend emerges starting in 1982 and accelerates from 1990. More striking, however, is the declining divergence in per capita GDP *within* coastal provinces and *within* interior provinces but the increasing divergence *between* coastal and interior provinces throughout the reform period.

The convergence pattern evident for GDP per capita in the early part of reforms is *not* observed in rural incomes. Rural incomes diverged persistently during the period, accelerating in 1990. Urban incomes also diverged, with some tapering off visible in 1994–95.

Trends in per capita GDP disparities are sensitive to the treatment of Beijing, Shanghai, and Tianjin . . .

Standard deviation of log GDP per capita

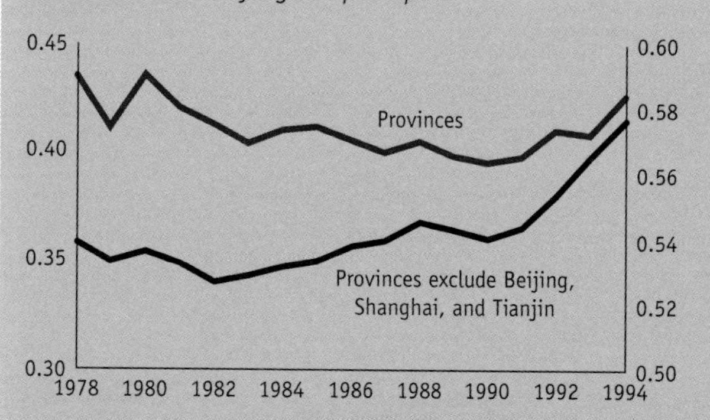

. . . but increasing divergence between the coast and the interior is unmistakable

Standard deviation of log GDP per capita

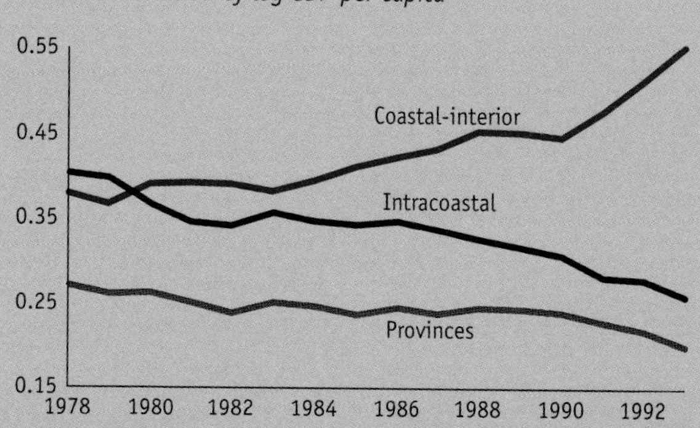

The coastal-interior divide is wider for rural incomes . . .

Standard deviation of log rural income

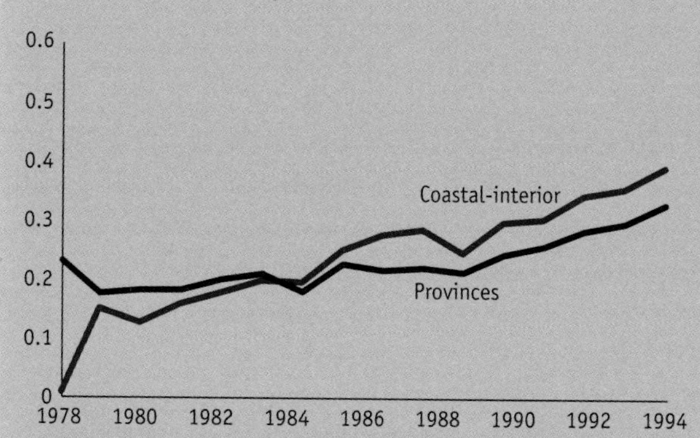

. . . than for urban incomes

Standard deviation of log urban income

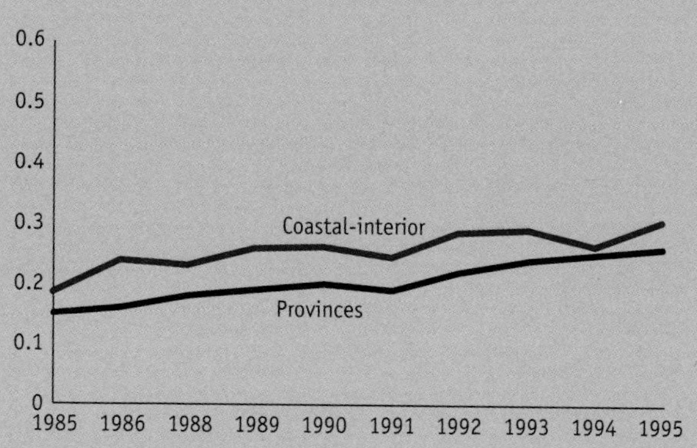

Note: See figure notes at end of chapter.
Source: China Statistical Yearbook, various years.

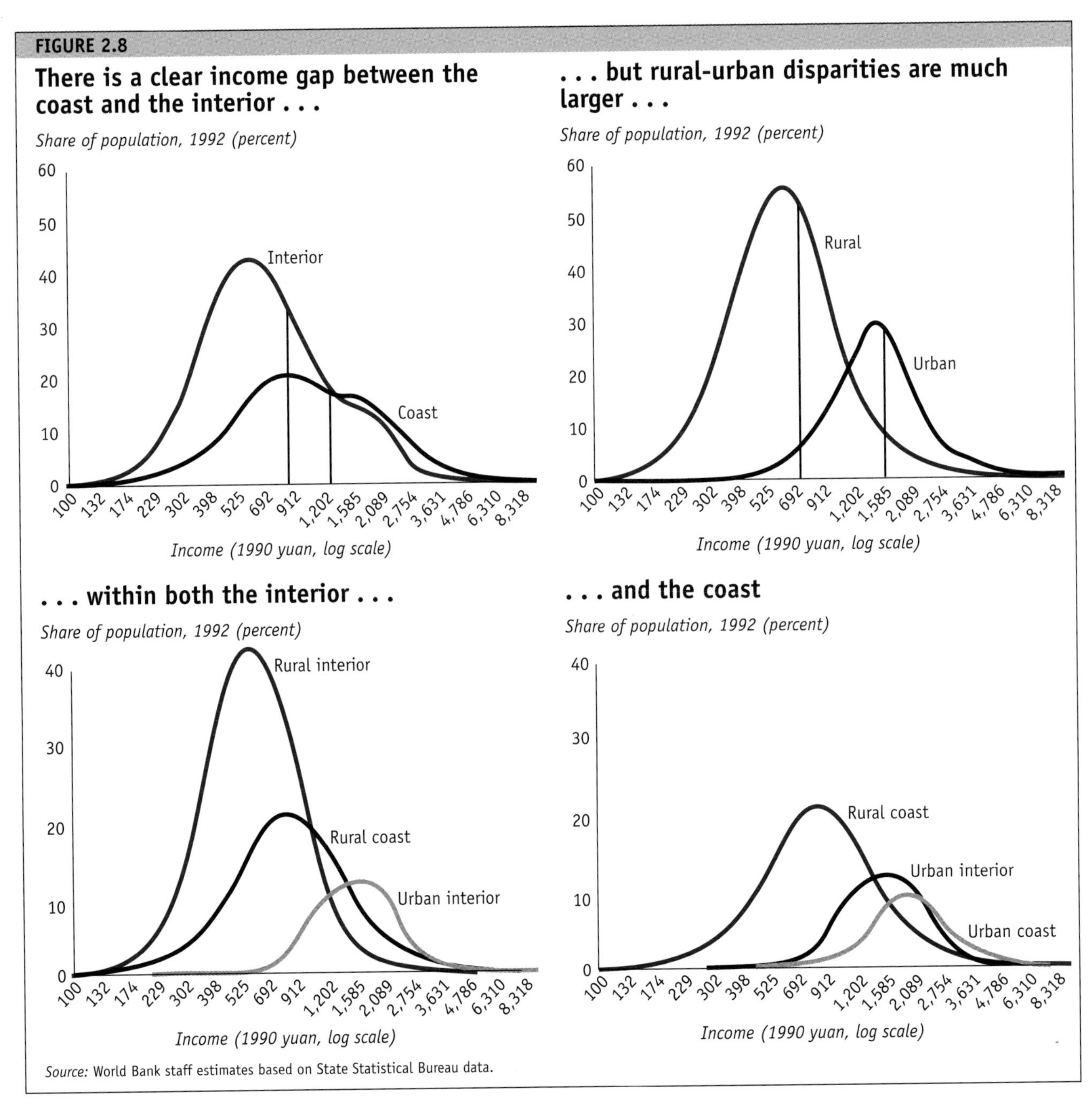

Source: World Bank staff estimates based on State Statistical Bureau data.

as two-thirds of total inequality remains *within* provincial borders.[10]

Provincial distributions for 1992 (the only year for which data are available) are telling (figure 2.8). They show that the income gap between the coast and the interior is significant; in 1992 average incomes in coastal China were 50 percent higher than in interior provinces. But in the same year the rural-urban income gap was twice as large.[11]

Provincial income disparities are increasing for several reasons. Since the start of reforms, coastal provinces have grown faster—and at an accelerating pace—than interior ones, fueling disparities in personal incomes. Coastal provinces grew 2.2 percentage points faster during 1978–94, 2.8 percentage points faster during 1985–94, and a remarkable 5 percentage points faster during 1990–94. Initial conditions, natural endowments, and preferential policies have combined to give coastal provinces a boost over interior ones in taking advantage of the opportunities created by reforms.[12]

First, the interior lags the coast in human capital development. Even before reforms, education and health levels were higher on the coast; the gap has since widened. Literacy, school attendance, and infant mortality are all

FIGURE 2.9

Access to health and education services was still widespread in the late 1970s and the 1980s . . .

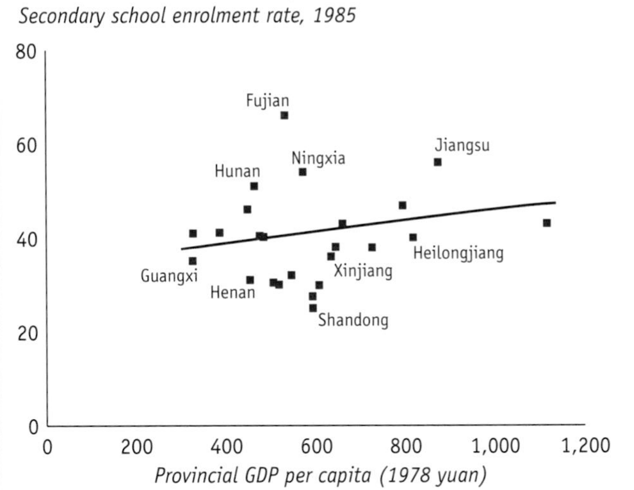

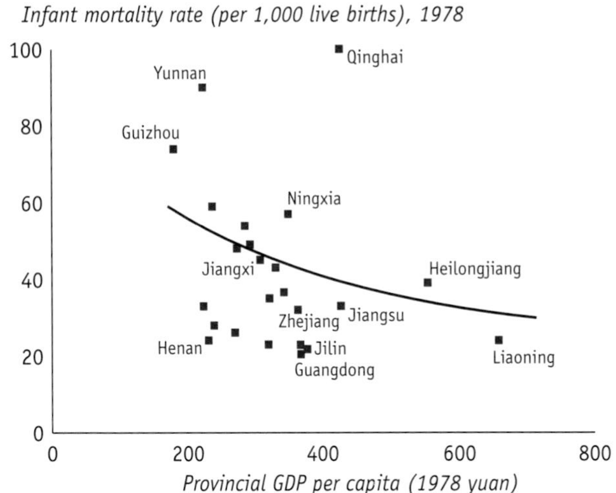

. . . but in the 1990s has become more dependent on income . . .

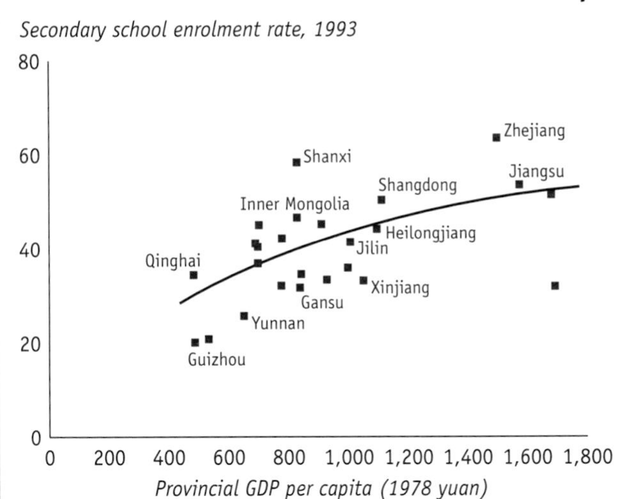

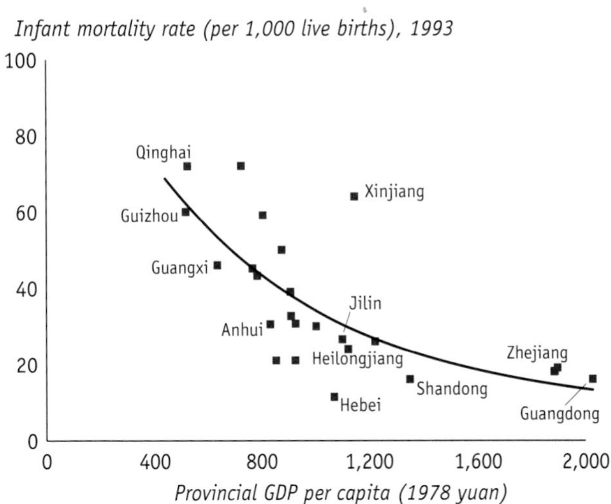

. . . reflecting the shrinking role of government, which itself has become less equal

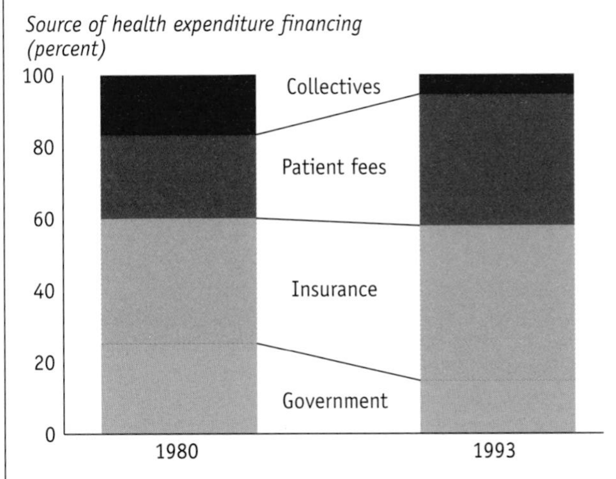

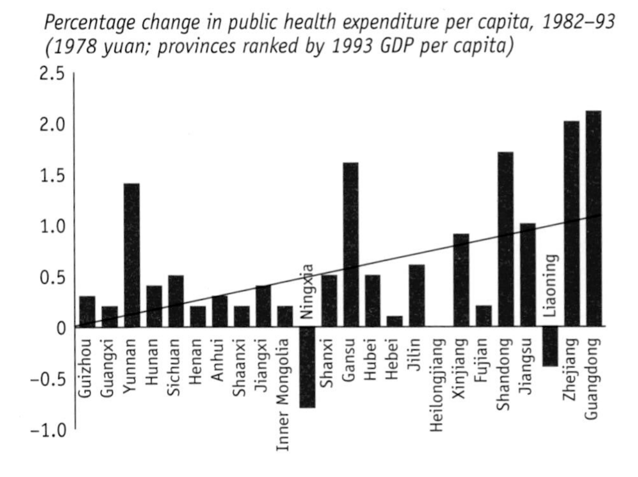

Note: Beijing, Shanghai, and Tianjin are excluded from the four scattergrams.
Source: China Statistical Yearbook, various years; China Ministry of Health data.

better on the coast. The presence of large ethnic minority populations in the interior likely contributes to this outcome. While primary school attendance is only slightly worse in the interior, high school attendance rates are significantly lower than on the coast.

Second, per capita investment levels on the coast are two and a half times those in the interior. And differences are not confined to levels of investment; they also extend to types of investment. In particular, the coast invests more in fast-growing industries, so more coastal residents work in industries that have seen high productivity gains. The coast also shows somewhat lower investment in state enterprises. More important, investment in fast-growing township and village enterprises accounts for nearly half of all non-state enterprise investment in most coastal regions, but less than a quarter in the interior. And these enterprises have grown faster on the coast than in the interior. Disparities in foreign direct investment and trade are even more striking. For example, in 1992 foreign direct investment in the coast accounted for more than 10 percent of total fixed investment; in the interior it accounted for less than 2 percent. And coastal regions received more than 85 percent of China's imports in 1993.

Third, as China has shifted from a closed and planned agricultural economy to an open, market-oriented industrial one, returns have increased to natural and geographical advantages. Natural advantages like harbors, transport corridors, proximity to world markets, and communication links have played a big role in spurring growth in coastal areas.

Fourth, regional policies have favored coastal areas by designating them for preferential treatment in foreign trade and investment. Credit has been allocated disproportionately to the coast, explaining in part investment differentials between the two regions. In addition, coastal provinces have often been the location of choice for pilot reform experiments.

Finally, decentralization of the fiscal system has fueled disparities in two ways (figure 2.9). It has increased the emphasis on cost recovery for social services, which reduces the poor's access to these services. It also has meant that richer provinces can spend more than poorer ones—for health, education, and infrastructure—further boosting their growth prospects. Furthermore, China's intergovernmental fiscal transfer scheme, which was equalizing in the 1980s, has not kept pace with rising regional disparities.

Notes

1. Deflating rural and urban incomes corrects for different increases in prices but does not adjust for differences in the level of prices between urban and rural areas in the initial year.

2. The State Statistical Bureau's urban household survey team estimated the monetary value of the main categories of in-kind income, including housing, health care, education, and pension contributions. The results increased the mean urban income in 1990 by some 80 percent. The bureau partly addressed the most important sources of underestimation of rural income in official data with the valuation of own-grain consumption at mixed prices, starting in 1990. Among the remaining areas requiring adjustment, an important one is imputed rent to reflect the value of owner-occupied housing. Based on household survey data for four provinces, Ravallion and Chen (1997) estimate this at about 6 percent of mean income. The changes in the rural-urban income ratio in this analysis are more pronounced than the ones found in Griffin and Zhao (1993). A 1988 household survey that corrected for a number of the concerns with official State Statistical Bureau data placed rural incomes at 41 percent of urban incomes and the national Gini coefficient at 38. The corresponding figures based on official data were 49 percent for the rural-urban income ratio and 33 for the Gini.

3. For urban incomes we rely on work undertaken by the State Statistical Bureau. For rural incomes household survey data from China's four southern provinces (Guangxi, Guizhou, Guangdong, and Yunnan) covering 1985–90 are used to gauge the effect data adjustments have on rural inequality; see Ravallion and Chen (1997). The multiyear nature of Ravillion and Chen's data set makes it possible to examine changes in inequality and their determinants. During 1985–90 the survey was longitudinal, returning to the same households over time. Ravallion and Chen used these results to construct panel data, which allow for analysis of the welfare of households over time. For more details on the data set and data problems of the State Statistical Bureau survey, see Ravallion and Chen (1997) and Jalan and Ravallion (1996b).

4. Other components of farm income also appear to have been undervalued, but this is less worrying because the shares of income involved are much smaller; 22 percent of rural incomes came from non-grain farm output, but only 10 percent of this was from own consumption.

5. To do this, Ravallion and Chen (1997) used a weighting diagram based on a food consumption bundle that ensures nutritional requirements are met, with an allowance for nonfood consumption anchored to the consumption behavior of the poorest 30 percent of the population. See Chen and Ravallion (1996) for full details on the methods for the alternative valuations and the new cost of living deflator.

6. There is Lorenz dominance between pre- and post-adjustment Lorenz curves for both 1985 and 1990, and between 1985 and 1990 for both unadjusted and adjusted curves.

7. Another problem has to do with the tabulation of urban data, starting in 1989. Until 1989 urban data published in the *China Statistical Yearbook* showed the share of households within a certain income range. Since 1989 these data have taken the form of average incomes corresponding to deciles of households, ranked by per capita income. But there is a serious problem with the way these data are tabulated. It appears that when county aggregates are processed by the provincial State Statistical Bureau teams, the decile (quintile)

means of all surveyed counties are averaged for each decile (quintile) to get the provincial mean in that decile (quintile). Differences in decile (quintile) boundaries from county to county are ignored, leading to significant underestimation of inequality. For 1989–95 this study uses urban data furnished directly by the State Statistical Bureau's urban household survey team in Beijing. Although these data have been properly tabulated and so do not have the problems of the published data, they come from a subsample of the whole survey. The urban survey covers 36,000 households but the data furnished to the study team come from a 17,000 household subsample. How the subsample was selected is not clear, but the mean incomes of this subsample are consistently higher than the published means for 1989–95. Thus changes in the urban income distribution between 1988 and 1989 should be treated with caution.

8. A standard rent per square meter was applied to average living space per capita based on the practices of the Housing Administration Authority for private housing rented privately.

9. The pension calculation understates the value of future pensions for urban workers in 1990 and to a lesser extent in 1995. Pension contributions represent only a small portion of future pension receipts. In the past the difference has been the obligation of enterprises, but this responsibility is now being assumed by local governments.

10. In 1992 mean income in the wealthiest province (Guangdong) was only 2.8 times mean income in the poorest province (Guizhou), whereas the ratio of the top to bottom income decile was 4.7 in the most equal province (Jiangxi) and as much as 16.0 in the most unequal province (Ningxia).

11. Rural residents in the interior have incomes that are much closer to coastal peasants than to urban residents in the interior. Similarly, people living in coastal cities have only mildly higher incomes than people living in interior cities but much higher incomes than people living in the coastal countryside (see figure 2.9).

12. This section draws on World Bank (1995b).

Figure notes

Figure 2.1 Adjustments are only indicative and reflect various corrections made to these data. Changes to rural incomes are based on corrections made by Ravallion and Chen (1997) to rural household data from four provinces (Guanqxi, Guizhou, Guangdong, and Yunnan) for 1985–90. Here we assume that these adjustments can be generalized to the country's rural areas as a whole. We also assume that changes made to the State Statistical Bureau's methodology in 1990 correct fully for the undervaluation of own-grain consumption. Thus adjusted data after 1990 reflect only the impact of regional price differentials and assume that the impact of these on rural inequality in 1990–95 remains as in 1990. Adjustments to urban data are based on information provided by the State Statistical Bureau's household survey team for 1990 and 1995. They reflect the inclusion of in-kind benefits on urban income. We assume that the level and distribution of pre-1990 in-kind benefits were as in 1990, and interpolate for the years between 1990 and 1995. Finally, we introduce a 15 percent cost of living differential between urban and rural areas, based on 1990 prices.

Box 2.1 figures Provinces include twenty-seven provinces and municipalities. Tibet, Hainan, and Qinghai are excluded because of incomplete data. Beijing, Shanghai, and Tianjin are incorporated in their neighboring provinces. The "new" provinces of Greater Hubei and Greater Jiangsu are classified as coastal.

chapter three

Understanding Inequality

The level of inequality and how it changes over time depend on people's employment, education, access to land, and gender, among other factors. China's transition from a centrally planned to a market economy is altering the economic landscape and with it the relationship between personal incomes and individual assets. Changes in the structure of employment are benefiting workers who can participate in the dynamic off-farm sector in rural China, or in the vigorous nonstate urban enterprises. A stronger link between productivity and wages is increasing the value of education. A more unified national market and increased openness to the outside world are rewarding people who live close to markets and have access to good infrastructure and good-quality land. The market is also placing a premium on male employees, whose time and mobility are not impeded by household or child-bearing responsibilities.

But other important changes are still to come. Continued reforms of the state will increasingly decentralize ownership and use of state assets. Changes in individuals' access to land, housing, and enterprise and other assets will profoundly affect future growth and income distribution. This process must be managed with caution, transparency, equity, and efficiency so that China can avoid the concentration of wealth that has stymied growth and development in other countries.

This chapter explores how employment, education, land, and gender are influencing inequality in China. The analysis is based on provincial economic indicators, summary urban household survey data on employment and educational status, and individual data for selected provinces and years.[1]

The structure of employment is changing

Employment and output have undergone substantial changes during the reform period. Township and village enterprises have transformed the face of rural China, while a flourishing nonstate sector has provided tremendous impetus to growth and productivity gains in urban areas. At the same time, in trying to cope with increased competition and less accommodating policies, state-owned enterprises are creating a new class of urban workers who are openly unemployed, underemployed (furloughed workers who remain on the payroll for reduced wages), or early retirees. These fundamental shifts in the ownership structure of production, output, and firm-level adjustments are changing the level and distribution of incomes. This section focuses on the links between incomes and employment. The analysis for urban areas suggests that ownership is important in determining the level of incomes but has limited influence on changes in incomes and inequality. In rural areas diversification into off-farm employment has induced inequality but also has improved the welfare of the poor.

The urban workforce goes nonstate

According to State Statistical Bureau household surveys, the share of urban employment in state-owned units has been growing since the late 1980s while that in collective enterprises has been shrinking (table 3.1).[2] The share of people working in private enterprises declined in the early 1990s but has picked up since 1992.

While the survey findings on changes in the employment shares of collective and private enterprises are consistent with national aggregate data, the trend in state employment shares is not (see table 3.1). This is most likely due to the many migrant workers in urban areas who are included in the aggregate employment data but are not captured in the household surveys. Surveys of migrant workers support this hypothesis. Because two-thirds of migrants go into the private sector and less than 15 percent obtain jobs with the state, including them in the urban surveys would generate a steeper increase in private employment and a probable decline for the state (see annex 1).

Both sets of data illustrate two well-known points. First, although the state continues to be the dominant employer in urban China, the private sector has

TABLE 3.1
Urban employment by ownership, 1989–95
(percent)

	State-owned units		Collective enterprises		Private and self enterprises	
Year	Sample mean	Aggregate data[a]	Sample mean	Aggregate data	Sample mean	Aggregate data
1989	75.9 (0.139)	70.8	19.4 (0.123)	24.3	4.7 (0.074)	4.9
1990	77.4 (0.133)	70.9	18.5 (0.124)	24.1	4.2 (0.053)	5.0
1991	79.9 (0.113)	70.2	16.1 (0.108)	23.8	4.0 (0.051)	6.1
1992	79.2 (0.110)	70.0	15.9 (0.103)	23.2	4.9 (0.056)	6.8
1993	79.2 (0.114)	69.8	15.5 (0.100)	21.3	5.3 (0.061)	8.9
1994	79.6 (0.121)	68.7	13.5 (0.104)	19.5	6.9 (0.068)	11.7
1995	79.7 (0.118)	67.1	13.2 (0.096)	18.1	7.1 (0.069)	14.8

Note: Numbers in parentheses are the standard deviation.
a. Includes employment in jointly owned and shareholding economic units.
Source: State Statistical Bureau data; *China Statistical Yearbook 1996.*

absorbed an increasing share of labor in recent years—a share that is even more pronounced if migrant workers are taken into account. Second, urban collective enterprises have become a less important alternative to the state as a source of jobs. National data show that employment in urban collective units dropped by more than 12 percent between 1989 and 1995. Except for construction, almost all sectors registered significant declines in employment; for example, manufacturing, which accounts for the largest portion of jobs in collective enterprises, lost nearly 20 percent of its labor force.

To determine how ownership affects income over time, a simple regression equation was estimated for each year between 1989 and 1995 (table 3.2). The results show that the shift from state and collective jobs to the private sector is associated with an increase in average incomes, reflecting the competitiveness and earnings potential of the growing private enterprises.[3] The increase in the coefficient of the state enterprise variable is curious, since it implies that the incomes of state employees relative to collective employees have been rising steadily since 1990. This might be due to differences in the operating environment. Collectives face strong competition from the private sector, so they offer wages in line with labor productivity and the market. State-owned units are still largely sheltered from market competition and continue to operate under a soft budget constraint.[4] Although ownership appears to be a significant determinant of the *level* of per capita urban incomes, the analysis suggests that it has limited power in explaining *changes* in urban incomes over time.[5]

TABLE 3.2

Determinants of urban income: Coefficients on ownership variables, 1989–95

	Share of working household members			
Year	State-owned units	Private enterprises	Number of observations	Adjusted R^2
1989	1.82 (12.66)	−1.40 (−4.82)	557	0.426
1990	1.34 (17.60)	1.70 (8.86)	580	0.556
1991	1.66 (17.68)	1.93 (9.70)	560	0.590
1992	1.94 (18.91)	2.75 (13.05)	580	0.632
1993	2.49 (19.05)	3.47 (13.84)	580	0.627
1994	2.64 (19.68)	2.96 (12.00)	580	0.588
1995	2.83 (20.50)	3.35 (12.36)	580	0.616

Note: The dependent variable is income per working household member. The omitted right hand variable is the share with collective jobs. Provincial dummy variables were included to control for province-specific elements. Numbers in parentheses are t-statistics.
Source: World Bank staff estimates based on State Statistical Bureau data.

To examine urban income inequality, its determinants, and how it changes, urban Gini coefficients for each province were regressed on the shares of employment by ownership and year dummy variables, controlling for province-specific fixed effects. The results suggest that ownership does not have a significant impact on urban income inequality. But the Gini coefficient used in the regression measures urban income inequality *within* each province. As a result the coefficient estimates only indicate the increase of urban income inequality within provinces over the past seven years. Increases in urban inequality at the national level due to rising disparities between provinces are not captured in this analysis.

The government has recently become concerned about differential wage increases across sectors. In part different increases suggest a welcome realignment in relative prices—an essential ingredient in transition. But different increases may also reflect the unfinished nature of transition if monopoly sectors and soft budget constraints are yielding wage rates that are not commensurate with productivity.

The rural workforce goes off-farm

The structure of rural employment shifted markedly between 1980 and 1995: farm incomes accounted for 60 percent of rural incomes in 1995, down from 78 percent in 1980 (top of figure 3.1). Increased opportunities for off-farm employment in the countryside have contributed significantly to growth and to inequality.

Overall, when rural workers diversify out of farm employment, they increase their incomes. Although off-farm incomes were sensitive to the growth cycles of overall rural incomes, the effects were less pronounced than for farm incomes. Per capita farm incomes grew by a whopping 17 percent a year between 1978 and 1984, but growth came to a virtual halt (0.6 percent a year) between 1985 and 1992. Off-farm incomes increased by 11 percent a year during 1978–84 and 3.3 percent a year during 1985–92 and created some opportunities for rural households to mitigate risk. But continued reliance on farming (particularly cropfarming) for most of their incomes meant that rural households remained vulnerable to slower agricultural productivity gains (middle of figure 3.1) and to adverse rural-urban terms of trade (bottom of figure 3.1). Grain yields stagnated between 1985 and 1989 while terms of trade turned

FIGURE 3.1

Workers become less vulnerable when they move off the farm

Diversification into off-farm employment has boosted rural incomes . . .

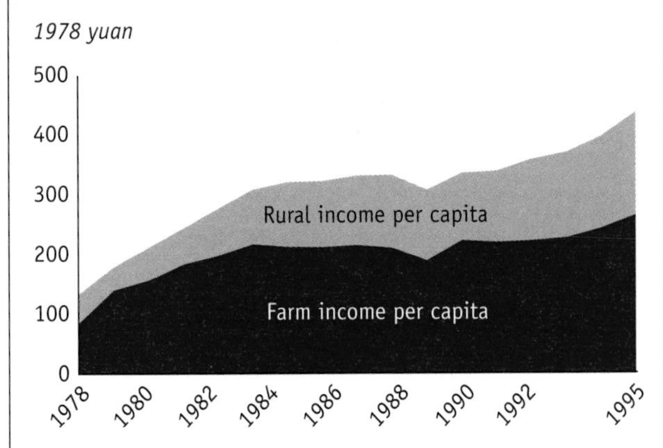

. . . but farm income growth depends heavily on gains in crop yields . . .

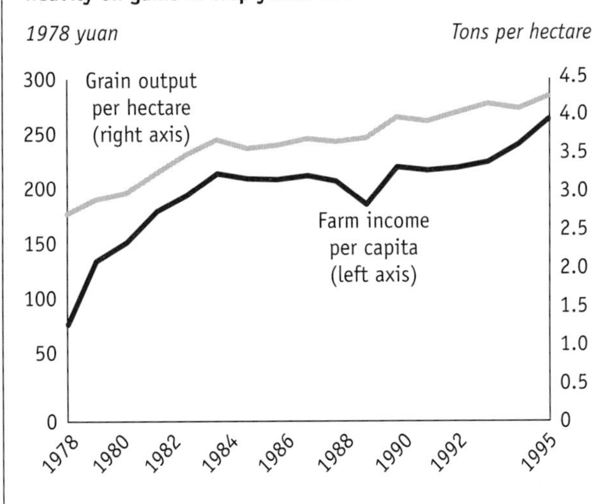

. . . and remains subject to adverse shifts in the rural-urban terms of trade

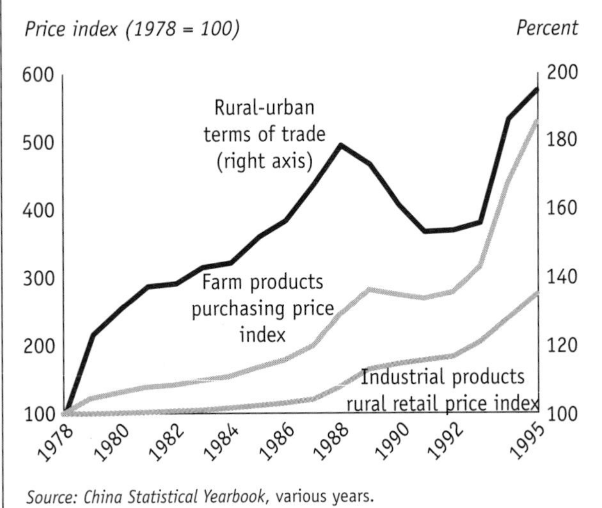

Source: China Statistical Yearbook, various years.

sharply against farmers between 1988 and 1991, driven by escalating input prices (as measured by the retail price index of industrial products in rural markets; see figure 3.1). As a result farm incomes stagnated during the second half of the 1980s and early 1990s.

When diversification is driven by higher returns from off-farm than from farm employment (pull factors) and entry barriers are high, off-farm employment may benefit mainly richer, more educated households and increase inequality, while having little impact on poverty. On the other hand, when diversification is largely a survival mechanism for poor households to supplement farm employment to meet their minimum needs (push factors), it can reduce both poverty and inequality.[6]

The four-province data set described in chapter 2 (Guangxi, Guizhou, Guangdong, and Yunnan) was used to examine the effect different sources of employment have on inequality in rural areas. Adjusting how income is measured does not alter the share of farm income in total income, but it does change the relative contributions of different sources within farm incomes because it generates a large increase in grain incomes (table 3.3). Grain accounts for 32 percent of rural incomes in the adjusted data compared with 21 percent in the unadjusted State Statistical Bureau data. Imputed rents for housing and durable goods ("other income") are included in the adjusted but not the unadjusted incomes; this reduces the proportion of off-farm income in the adjusted data.

The share each source of income contributed to inequality changed between 1985 and 1990.[7] Two trends are clear, regardless of how income is measured. First, by 1990 off-farm employment had become the largest source of inequality. Second, transfers were equalizing, since their contribution to inequality fell between 1985 and 1990 (figure 3.2). While both public and private transfers helped reduce inequality, private transfers were particularly effective, accounting for more than 10 percent of inequality in 1985 but less than 4 percent by 1990. Thus migration opportunities are clearly available to poor households in the south and play an important role in equalizing incomes.

The data adjustments lower overall income inequality in the rural south but increase the share of the reduced inequality attributable to the distribution of grain income. The contribution of grain income to total inequality increased 35 percent between 1985 and 1990 but remained small in the original State Statistical Bureau definition of income. By contrast, when adjusted incomes are used, the contribution of grain to inequality is larger initially and increases 60 percent (from 8.8 to 14.3 percent).

Overall, employment diversification *increased* inequality for the four provinces analyzed. Evidence from Sichuan and Jiangsu (Burgess 1997) corroborates this finding. However, the question remains whether diversification led to increasing inequality because it benefited mainly rich households or whether it also improved the welfare of the poor. Evidence from Sichuan and Jiangsu suggests that diversification has enhanced overall welfare in rural China.

Diversification has the potential to increase the incomes of the poor and can help stabilize incomes and mitigate risk.[8] Welfare can improve even with a constant level of income if there is a smoother flow. Because Chinese households have limited access to means of ex post consumption smoothing (for example, through credit markets and transfers), ex ante income smoothing, through occupational choice, may play an important role. Diversification can help mitigate risks associated with farm employment and make the lives of rural residents more secure.

Two measures are used to analyze the incidence of diversification: share of off-farm income in total income and probability of having at least one household member registered in an off-farm job as their primary source of employment. As expenditures per adult rise, off-farm income increases as a share of total income (table 3.4). The probability of having at least one household member engaged primarily in off-farm employment also increases with rising welfare. These findings confirm that increasing off-farm income is the main factor driving rising living standards in China. These patterns are more pronounced for Jiangsu than Sichuan.

TABLE 3.3

Average income by source in the rural south, 1985–90

(percentage of six-year mean income)

Source	Unadjusted (original) income	Adjusted (revised) income
Farm income	72.3	71.9
Grain	21.1	31.5
Other farm	51.1	40.5
Off-farm income	25.8	19.9
Transfers	5.1	4.5
Other income	1.4	7.8
Joint costs	−5.2	−4.1
Total	100.0	100.0

Note: Unadjusted incomes are from the State Statistical Bureau. *Adjusted incomes* reflect valuation and deflator adjustments. *Other farm income* includes remuneration from nongrain crops, animal husbandry, forestry, and fisheries. *Off-farm income* is income generated from collectively or individually owned businesses and state employment. *Transfers* include both public and private ones; private transfers are largely remittances from migrant household members. *Other income* covers other factor incomes and imputed rent on housing and durable goods for the adjusted definition of income. *Joint costs* are costs that cannot be apportioned between factor income components.

Source: State Statistical Bureau data; Ravallion and Chen 1997.

FIGURE 3.2

The contribution of off-farm income to inequality is large and rising while transfers are increasingly equalizing, 1985 and 1990

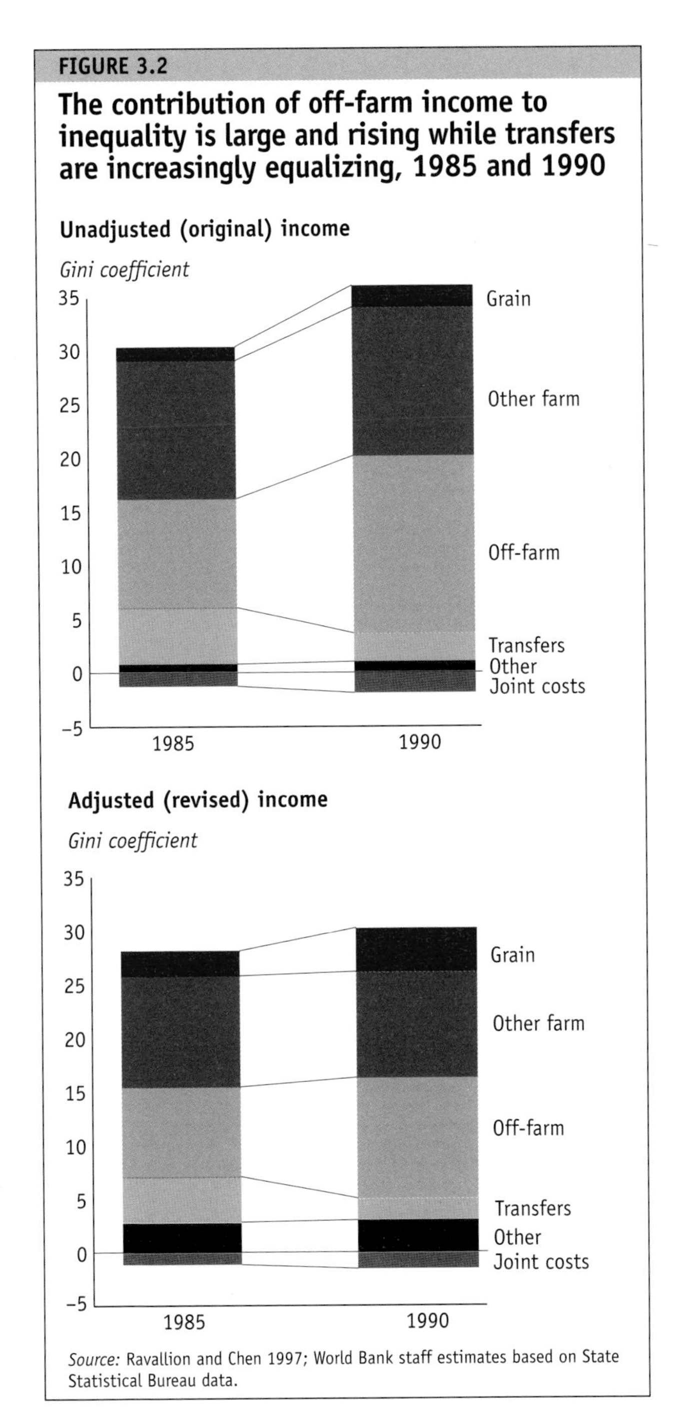

Source: Ravallion and Chen 1997; World Bank staff estimates based on State Statistical Bureau data.

TABLE 3.4
Incidence of diversification in rural China
(percent)

House hold ranking by expenditure per equivalent adult decile[a]	Off-farm income as a share of total income		Share of households with at least one member registered in an off-farm job	
	Rural Sichuan	Rural Jiangsu	Rural Sichuan	Rural Jiangsu
1	11.0	14.9	11.5	24.1
2	10.8	16.5	13.2	24.7
3	12.4	23.7	14.1	41.2
4	13.3	24.8	19.0	43.4
5	14.9	26.1	23.0	53.1
6	14.7	29.5	21.4	59.5
7	16.3	31.0	24.2	60.4
8	18.8	35.4	30.1	67.7
9	20.3	37.2	35.3	68.1
10	24.6	45.5	42.6	75.3
All	15.7	28.5	23.4	51.8

a. The use of an equivalence scale as a deflator is important in this context because initial tabulations using per capita expenditure as the sorting variable showed a marked decrease in family size with increasing per capita expenditure. In China, where the basic earning asset (land) has a relatively egalitarian distribution and where the land allocation process takes into account the needs of children (though with a lower weight than that of adults), larger families with a higher proportion of children de facto appear poorer. Despite the tenuous assumption on which equivalence scales are based, they are better than per capita measures in this setting.
Source: Burgess 1997.

But even the poorest households obtain sizable shares of their income from off-farm employment, and the probability of having at least one household member employed primarily off-farm is not negligible for households in the lower deciles. Thus the benefits of diversification appear to have reached low-income groups in rural China. Taking into account the enhanced ability of diversified households to mitigate income risks, diversification in rural China has increased inequality but also improved welfare.

The value of education is increasing

Throughout their long history, the Chinese have valued education. But during much of the 1980s "head and body were upside down." Manual laborers were likely to earn as much as, if not more than, doctors or engineers who had devoted years to study. Before reforms the returns to education were notoriously low, but increased market orientation has been correcting this anomaly, bringing China closer to international norms. This adjustment is affecting income distribution patterns and increasing the demand for schooling.

Education affects income inequality in two ways. First, as the value of education increases, income inequality tends to rise: manual laborers no longer earn as much as doctors, creating an income gap. Second, changes in the distribution of educational attainment may aggravate or attenuate disparities in future income. If all children have the possibility of obtaining a university degree, regardless of whether their parents are laborers or doctors, income disparities among the next generation may decline. But without equal access to education, income disparities are likely to intensify.

The analysis confirms increased returns to education in both urban and rural areas over the reform period. Evidence on the distribution of educational attainment points to greater equality for primary and middle school education but increased inequality for higher schooling. Overall, changes in educational attainment and its returns have contributed to increasing inequality in China.

Urban China embraces education

Urban Chinese are becoming more educated, but in an increasingly unequal fashion. At the same time, markets are bidding up the value of education.[9] Combined, these developments suggest increasing divergence in labor income and explain much of the increased income inequality in urban areas.

First, educational attainment in urban areas has improved since the late 1980s (table 3.5).[10] Between 1988 and 1995 the share of working household members with more than a high school education increased substantially—by as much as 70 percent for those with university degrees and above. This jump reflects increasing demand for skilled labor as China's industrial structure changes and markets mature. Second, the data show small and narrowing dispersion in the attainment of junior middle school education, in line with the government's policy of universal basic education. But disparities are larger for higher education, especially at the post-secondary level. National data confirm trends identified in the survey of urban workers, showing that the population as a whole is becoming more educated, but post-secondary education is increasingly dispersed (table 3.6).

Third, regression analysis shows that urban income increases with educational attainment (table 3.7). Fourth, estimated schooling coefficients for each year show increasing returns to higher education over the period.

Rural China—schooling matters

There are large disparities in educational attainment between rural and urban areas. A 1988 national survey found that urban residents had 9.6 years of education compared with 5.5 years for rural residents (Knight and Li 1996).[11] Moreover, this gap has narrowed only slightly over four decades. Thus disparities in educational attainment may help explain the large differences in the *level* of rural and urban incomes (see chapter 2). But they are unlikely to explain changes in the ratio of rural to urban incomes. One possible explanation for the widening income gap is that returns to education have increased faster in urban areas than in rural areas because the bulk of the population remains employed in farming, where educational attainment continues to have a limited impact on incomes.

TABLE 3.5

Urban educational attainment, 1988, 1991, and 1995

(sample mean, percent)

Year	Primary and below	Junior high and above	High school and above	Post-secondary
1988	14.3 (10.0)	85.7 (10.0)	46.4 (11.8)	11.2 (5.8)
1991	9.5 (6.2)	90.5 (6.2)	55.5 (11.2)	15.5 (7.8)
1995	10.2 (5.6)	89.8 (5.6)	59.1 (12.7)	19.0 (9.5)

Note: Numbers in parentheses are the standard deviation. Tabulated survey data provide information on mean household income per capita for six income groups (top and bottom decile and four quintiles in between), with corresponding household size and number of working household members, broken down by level of education for all provinces for three years.

Source: State Statistical Bureau urban household survey team.

TABLE 3.6

National educational attainment, 1981 and 1993

(mean, percent)

Year	Uneducated	Primary	Secondary	Post-secondary
1981	40.9 (12.3)	33.0 (6.0)	25.2 (9.6)	0.9 (1.0)
1993	21.4 (11.4)	40.1 (7.5)	36.6 (10.4)	2.0 (2.4)

Note: Numbers in parentheses are the standard deviation. Table data show the ratio of people with corresponding educational attainment to the population 6 years old and above.

Source: China Statistical Yearbook, various years; *China Population Yearbook 1994.*

TABLE 3.7

Determinants of urban income: Coefficients on education variables, 1988, 1991, and 1995

	Shares of people with schooling level				
Year	Junior high	High school	Post-secondary	Number of observations	Adjusted R^2
1988	−1.60	0.80	4.61	173	0.61
1991	1.22	3.12	5.13	174	0.73
1995	1.22	2.30	5.44	174	0.83

Note: All estimates are statistically significant at the 5 percent level. The dependent variable is income per employee. The omitted variable is the share who received education up to primary level. Province dummy variables were included to control for fixed effects.

Source: World Bank staff estimates based on State Statistical Bureau data.

TABLE 3.8
Rural southern educational attainment, 1985 and 1990
(mean, percent)

Year	Primary	Junior high	High school	Technical school	University
1985	0.34 (1.15)	0.43 (1.20)	0.17 (0.91)	0.01 (0.22)	0.001 (0.09)
1990	0.32 (1.09)	0.48 (1.17)	0.16 (0.85)	0.01 (0.23)	0.002 (0.11)

Note: Numbers in parentheses are the standard deviation.
Source: World Bank staff estimates based on State Statistical Bureau data.

Still, the returns to education are also likely to have risen in rural areas because of the increasing importance of off-farm employment since the start of reforms. Nonfarm activities accounted for 33 percent of rural incomes in 1994 and 22 percent in 1985, up from 7 percent in 1978. There is clear evidence from Sichuan and Jiangsu that better-educated households are more able to obtain off-farm employment. Moreover, the greater is the share of off-farm income in the total, the higher is the level of income (see also table 3.4). This finding suggests that the distribution of educational attainment helps explain the distribution of income. The effect of education on access to off-farm employment is more pronounced in Jiangsu than in Sichuan. This finding implies that more education is required to enter off-farm activities in Jiangsu, probably reflecting the more specialized nature of these activities in this more developed province.

Using the four-province rural data set described earlier (Guangxi, Guizhou, Guangdong, and Yunnan), it is possible to examine changes over time in the distribution of and returns to education and the contribution of educational attainment to income inequality. The sample data show that the standard deviation of educational attainment *decreased* for primary, middle, and high school but *increased* for technical and university education (table 3.8). The evolution of returns to education was captured in a regression analysis undertaken to examine the importance of various assets in determining real per capita incomes (table 3.9). For both the State Statistical Bureau's original income data and the adjusted incomes, the results show that returns to schooling (except university) rose during 1985–90.

What do these findings on educational attainment and its returns imply for the contribution of schooling to levels of and changes in rural income inequality?[12] Several conclusions emerge from the analysis (table 3.10 and figure 3.3). First, the income determinant variables included in the regressions explain only a small portion of income inequality in any one year. The unexplained (residual) component of the variance in incomes still accounts for 70–80 percent of the level of inequality and one-half to two-thirds of the increase in inequality (as measured by the Gini coefficient). Second, all the education variables combined account for only 2.5–3.0 percent of income inequality in any given year, but they explain about 8 percent of the increase in inequality over the period. Third, primary education reduced inequality, while other levels of education increased it, although the contribution of each was small in all cases (and negligible in the case of university education). Because returns to education increased for all the education variables (except university), and since primary schooling (as the highest level of schooling) is negatively correlated with income, the higher returns to primary education reduced inequality. By contrast, the large increases in the returns to higher education contributed to inequality, although this effect was dampened by an improvement in the distribution of secondary education; both the standard deviation of secondary education and its correlation with income fell between 1985 and 1990. Thus a more equal distribution of secondary schooling helped attenuate the effect its higher rate of return had on overall income inequality.

Land remains a powerful source of social protection

Land is probably the most important asset to which rural households have access; in China it accounts for 59 percent of rural wealth.[13] Thus the distribution of land affects income inequality and welfare in rural China. Elsewhere, the distribution of land has also been found to be an important determinant of growth.

The relationship between growth and inequality is of longstanding interest among economists. Deininger and Squire (1996) find evidence of a negative but weak relationship between initial *income* distribution and future growth but a strong relationship between initial *land* distribution and growth (figure 3.4). The authors suggest that this difference may reflect the importance of land in the ability of the poor to access credit markets and make productive investments. In China an alternative explanation is fiscal: access to land is an effective form of social protection in rural China and eliminates the need for costly public transfers financed from general tax revenues.

The 1988 national survey noted earlier found that, unlike in other developing countries, there is little incidence of landlessness in China: almost all rural households "own" at least some land (McKinley and Griffin 1993). Also unlike other developing countries, landlessness is not a significant determinant of poverty in China. The incidence of poverty is higher among the landless, but not dramatically so, and most of the landless are not poor.[14]

McKinley and Griffin (1993) calculate the Gini coefficient for three distributions of land—physical units, irrigation-adjusted units, and land values. The Gini coefficient for the distribution of physical units of land is 54.3. This figure is probably the most comparable to Gini coefficients reported for other countries and compares favorably with them.[15] More important for the effect of land distribution on rural welfare,

TABLE 3.9

Determinants of rural southern income, 1985 and 1990

	Regression coefficient				Correlation coefficient with total			
	Unadjusted (original) income		Adjusted (revised) income		Unadjusted (original) income		Adjusted (revised) income	
Variable	1985	1990	1985	1990	1985	1990	1985	1990
Intercept	333.02	420.33	401.91	463.38				
	(23.52)	(23.99)	(27.60)	(29.82)				
Fixed productive assets per capita	0.32	0.18	0.36	0.21	0.24	0.18	0.27	0.23
	(24.36)	(17.01)	(26.35)	(19.47)				
Household size	-14.96	-21.30	-16.48	-24.08	-0.04	-0.05	-0.05	-0.05
	(-12.33)	(-14.38)	(-13.21)	(-18.33)				
Household labor force per capita	155.69	126.55	181.50	147.89	0.20	0.15	0.21	0.19
	(12.98)	(9.91)	(14.71)	(13.05)				
Hilly area	-74.48	-101.67	-93.30	-89.49	0.05	0.08	0.04	0.09
	(-12.22)	(-13.42)	(-14.88)	(-13.32)				
Mountainous area	-144.39	-201.73	-175.19	-191.74	-0.24	-0.29	-0.27	-0.30
	(-24.48)	(-29.43)	(-28.88)	(-31.57)				
Owned cultivated land area per capita	0.14	0.13	0.17	0.33	0.08	0.05	0.09	0.13
	(6.90)	(3.77)	(8.41)	(10.89)				
Area of hilly land per capita	-0.01	-0.002	-0.004	0.01	0.00	-0.05	0.00	-0.03
	(-1.45)	(-0.15)	(-0.92)	(0.80)				
Area of fishpond land per capita	0.08	1.27	0.07	0.89	0.06	0.16	0.05	0.13
	(2.96)	(14.40)	(2.53)	(11.40)				
Primary school	38.0	50.9	36.6	49.6	-0.12	-0.13	-0.12	-0.12
	(3.87)	(3.93)	(3.63)	(4.32)				
Middle school	78.2	106.9	76.5	97.5	0.06	0.05	0.06	0.05
	(7.93)	(8.30)	(7.55)	(8.53)				
High school	117.6	171.6	115.6	148.2	0.13	0.13	0.13	0.10
	(10.75)	(12.13)	(10.27)	(11.81)				
Technical school	133.4	216.6	121.1	193.1	0.03	0.05	0.03	0.05
	(4.89)	(7.55)	(4.32)	(7.59)				
University	226.6	253.7	213.3	203.8	0.01	0.03	0.01	0.03
	(3.43)	(4.43)	(3.14)	(4.02)				

Note: The dependent variable is income per capita, in constant prices. The education variables are all dummies for the highest level of education reached by the household workforce, with the omitted dummy variable being for an illiterate. *Fixed productive assets* is the survey valuation of all immobile productive farm assets, expressed in constant prices and normalized by household size. *Labor force* is the number of able-bodied workers. Land variables include *cultivated land, hilly land,* and *fishpond* as areas of land owned per person in the household, and *hilly area* and *mountainous area* as dummy variables for the geographic area in which the household lives, with the omitted dummy variable being that for households living on the plains. Numbers in parentheses are t-statistics.
Source: Ravallion and Chen 1997.

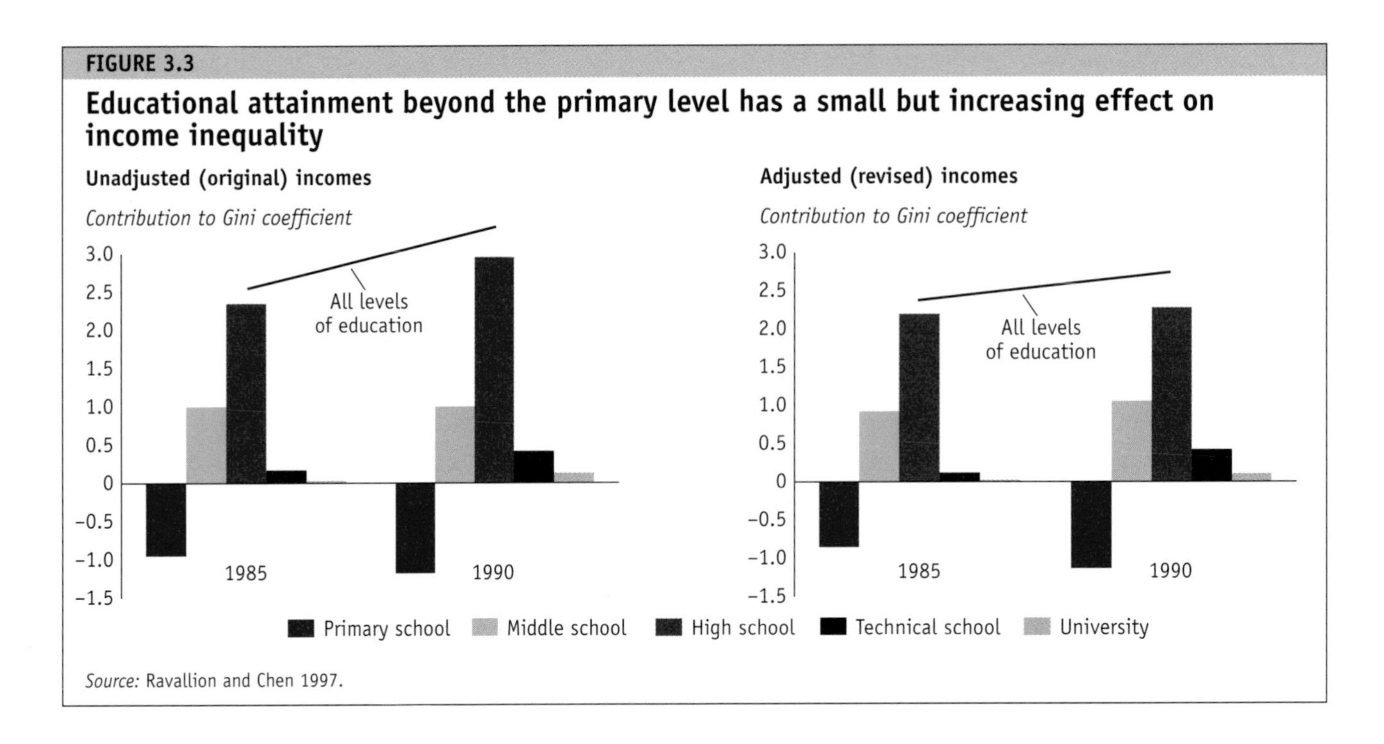

FIGURE 3.3

Educational attainment beyond the primary level has a small but increasing effect on income inequality

Source: Ravallion and Chen 1997.

TABLE 3.10

Contribution of income determinants to inequality in the rural south, 1985 and 1990

	Unadjusted (original) incomes			Adjusted (revised) incomes		
Variable	1985	1990	1985–90 Gini coefficient	1985	1990	1985–90 Gini coefficient
Productive assets	5.86	3.45	−11.24	6.68	4.34	−33.82
Household size	0.21	0.36	1.25	0.25	0.63	6.81
Labor	2.42	1.41	−4.74	2.91	2.30	−7.61
Hilly area	−0.69	−1.39	−5.67	−0.80	−1.49	−12.88
Mountainous area	7.31	10.96	33.13	9.47	11.95	52.40
Cultivated land	0.53	0.19	−1.86	0.74	1.32	10.85
Hilly land	0.00	0.01	0.08	0.00	−0.02	−0.35
Fishpond	0.12	2.17	14.67	0.09	1.40	22.79
Primary school	−0.94	−1.17	−2.59	−0.85	−1.13	−5.59
Middle school	1.00	1.00	1.04	0.92	1.05	3.15
High school	2.36	2.96	6.60	2.20	2.28	3.60
Technical school	0.17	0.42	2.00	0.12	0.42	5.19
University	0.04	0.14	0.70	0.03	0.11	1.34
Residual	81.61	79.49	66.65	78.23	76.84	54.13
Total	100.00	100.00	100.00	100.00	100.00	100.00

Source: Derived from table 3.9.

however, is that adjustments for quality yield significantly lower inequality. The Gini coefficient for irrigation-adjusted units, for example, is 50.9. When the measurement is based on land values rather than physical units the Gini drops to just 31.0—lower than the Gini that the authors found for income distribution. This low inequality of land distribution in China has important implications for income inequality and welfare of the poor. [16]

The four-province rural southern data set is used to examine how land affects income inequality over time. As expected, all land variables combined contributed little to income inequality in 1985—7.0 percent using unadjusted (original) income and 9.5 percent using adjusted (revised) income (figure 3.5). Most of this contribution is due to the gap between lower incomes earned by people living in mountainous rural areas relative to people living on the plains—a gap that increased between 1985 and 1990. Indeed, the distribution of households between mountainous areas and the plains accounted for more than half the increase in the Gini coefficient using adjusted incomes (one-third using unadjusted incomes). The income gap is partly due to returns on farm labor, which differ by location. But it may also reflect that there is less off-farm employment in mountainous areas.[17] Access to farm land (cultivated land per capita) also appears to be an important source of higher inequality over time for adjusted incomes, accounting for about 10 percent of the increase. This is so despite a decline in the standard deviation of cultivated land distribution, because adjusted incomes show increasing returns to land and a higher correlation coefficient between cultivated land and incomes.[18]

China's egalitarian land distribution has almost erased the hunger and malnutrition that are prevalent in other low-income countries. Thus the distribution of land can be thought of as a decentralized form of social protection. Unlike in other developing countries, institutional features in rural China have ensured that even the poor have access to sufficient land to meet the bulk of their caloric needs (box 3.1). Although household income inequality increased during the 1980s, undernutrition did not. Instead, key institutional features continued to protect against nutritional risk and were perhaps even strengthened by incentives for improved production.

Several features contribute to the sustainability of China's land distribution. In a large, geographically diverse economy with limited infrastructure (such as China), social protection through the distribution of land has many advantages over in-kind or cash transfers. Land distribution empowers households to avoid undernutrition rather than depend on government and other institutions. This form of social protection is also compatible with the profit motive because most of the bene-

FIGURE 3.4

Land equality is good for growth

GDP growth, 1960–92

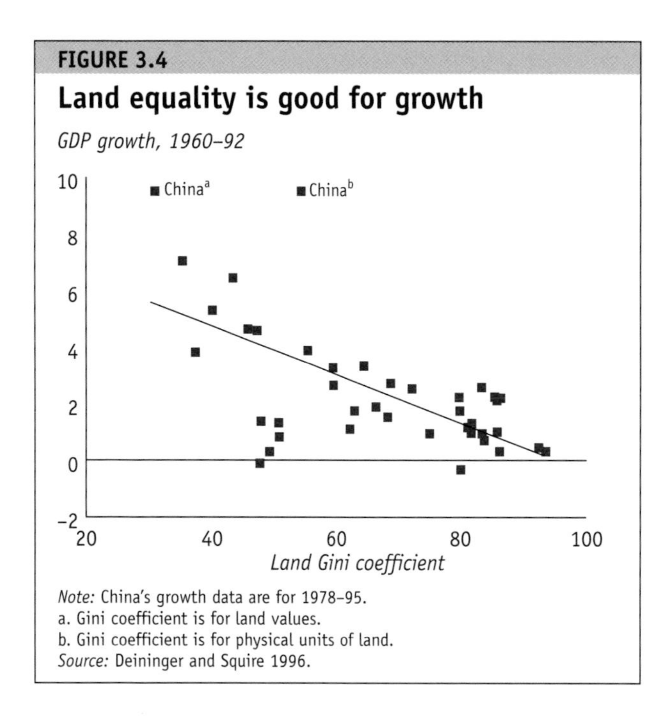

Note: China's growth data are for 1978–95.
a. Gini coefficient is for land values.
b. Gini coefficient is for physical units of land.
Source: Deininger and Squire 1996.

FIGURE 3.5

The contribution of land to income inequality reflects the growing income gap between people living in the plains and those living in the mountains

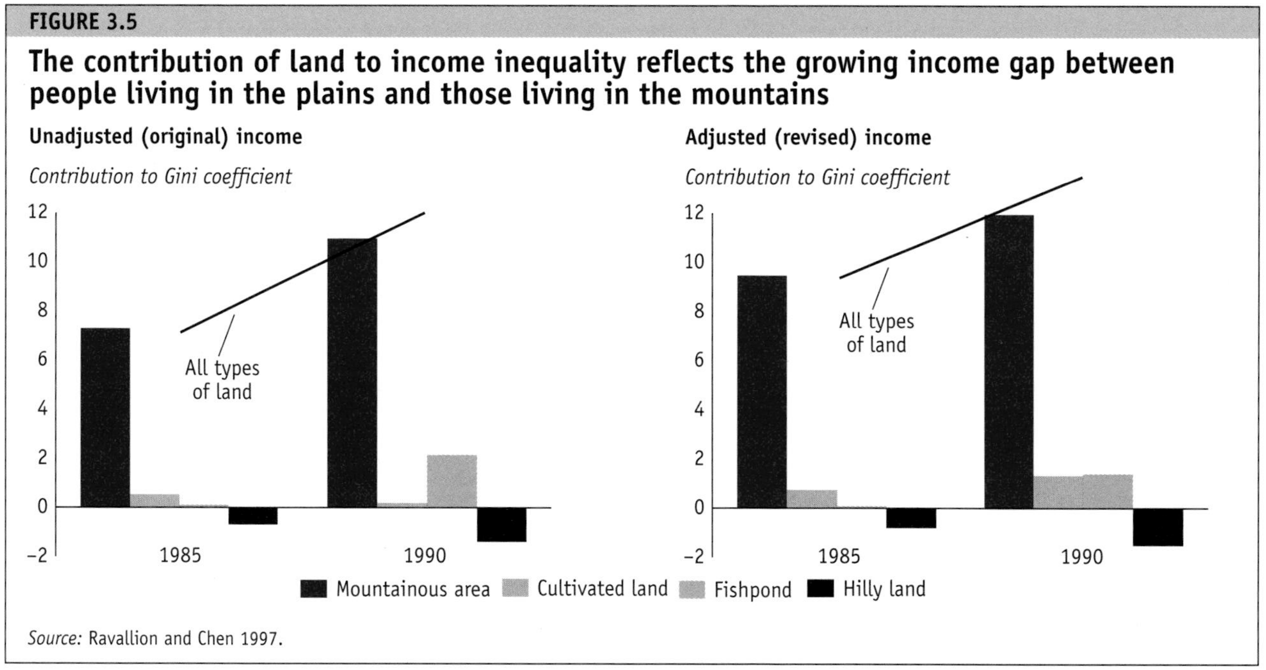

Source: Ravallion and Chen 1997.

BOX 3.1

Distribution of land in China

In the 1950s extensive land reforms provided peasants with sufficient land to meet their nutritional needs. During the era of collective production, 1950–78, when monetization and diversification of the rural economy were limited, access to arable land, food production, and nutritional and economic welfare were closely correlated. Given the limited importance of currency, living standards were typically assessed with respect to the grain sufficiency of households. The 1959–61 famine led to the household responsibility system; village land was divided among households, which had control over its outputs and profits. The land allocation rules implemented as part of the household responsibility system favored universal access to and an egalitarian distribution of land. Although village governments were autonomous in their decisions, they tended to choose egalitarian rules based on household demographic composition because villagers were accustomed to egalitarian treatment.

Source: Burgess and Murthi 1996.

fits of land access accrue to the household that maintains it. Moreover, the administrative costs of operating the system have been minimal because land redistribution was historically achieved in a single step. Finally, macroeconomic fluctuations do not affect the level of benefits because the system does not depend on fiscal revenues.

Most policies to improve nutritional status focus on transferring income or food. A third alternative—providing rural people with opportunities to produce food—appears to have been quite successful in China, where high nutritional status has been obtained without resorting to large-scale direct nutrition interventions. Because growth has also benefited the poor, it has improved their nutritional welfare. But the pattern of growth and possibly growth rates may have been less favorable in the absence of a relatively even distribution of assets in rural China.[19] Although the current distribution of land may not be economically optimal, the equity and social protection features of the system serve important functions. Alternatives to this system should explicitly consider measures to compensate those who would be made nutritionally vulnerable by its absence.

Women are increasingly at risk

The status of Chinese women compares favorably with that of women in other Asian countries. Chinese girls are more likely to attend primary school, and women are more likely to have paid employment. Yet alongside persistent problems such as inadequate access to social services for women in poor areas, new problems are emerging with deeper economic reforms. In fact, unfettered market orientation is threatening past achievements in gender equality and reinforcing a cultural predisposition toward differential treatment of men and women. For example, if reproduction and child care are exclusively female responsibilities, women will earn less in the labor market and parents will have less incentive to educate girls because expected returns will be lower than for boys. In turn, less-educated women will earn less in the future. Thus income inequality between men and women is exacerbated by gender bias in household decisions about investment in children and by occupational segregation and wage discrimination in the labor market.

Here we use household data from Sichuan and Jiangsu to test for gender discrimination at the household level by examining gender bias in the household allocation of food, calories, health, and education, since these expenditures influence children's survival probability and the welfare outcomes of surviving children; and selective abortion of female fetuses, leading to skewed sex ratios at birth (Burgess and Zhang 1996). Three key results emerge from the analysis of household spending on food, education, and health. First, there is no evidence of gender bias in the allocation of food and calories, suggesting that parents feel unconstrained in this area thanks to universal access to and egalitarian distribution of land. Second, there is evidence of a significant bias against girls (0–4 age group) in the allocation of health goods in the poorer, less diversified province (Sichuan), but not in the richer, more diversified one (Jiangsu).[20] Third, overall investments in human capital are higher for boys than girls, with both provinces showing a pro-male bias in expenditures for secondary education. Taken together, the health and education results suggest that discrimination tends to focus on expenditures when parents have to make discrete and costly decisions regarding investment in their children.

Income growth and diversification appear to erode gender discrimination. This result is obtained when the rural samples for both provinces are split into poorer and richer subsamples and the analysis is extended to

TABLE 3.11
Mortality rates and sex ratios in Sichuan and Jiangsu, 1990

	Rural Sichuan			Urban Sichuan			Rural Jiangsu			Urban Jiangsu		
	Mortality rate[a]		Sex	Mortality rate[a]		Sex	Mortality rate[a]		Sex	Mortality rate[a]		Sex
Age	Male	Female	ratio[b]	Male	Female	ratio[b]	Male	Female	ratio[b]	Male	Female	ratio[b]
At birth												
Reverse survival method[c]			111.9			110.2			115.5			112.2
Actual			115.8			110.2			120.8			116.0
0	25.1	30.0	112.6	29.2	27.8	107.7	15.3	15.5	115.5	12.9	12.6	112.2
1	4.3	5.5	113.7	3.6	3.6	107.2	3.1	2.8	113.7	1.8	2.0	110.8
2	2.7	3.2	111.3	2.6	2.5	105.8	2.5	2.1	111.3	1.6	1.3	108.8
3	1.5	1.7	111.5	1.4	1.5	106.5	1.4	1.3	110.0	1.0	0.8	106.8
4	1.0	1.0	110.4	0.9	0.9	106.2	1.1	0.9	109.3	0.8	0.9	106.8
0–4	10.3	11.8	111.8	7.0	6.7	106.6	5.1	4.8	112.1	3.8	3.6	109.1
5–9	1.2	1.0	109.4	0.9	0.5	106.4	0.7	0.5	108.3	0.6	0.4	106.0
10–14	1.0	0.7	106.4	0.8	0.5	105.4	0.4	0.4	106.6	0.4	0.3	106.8
15–19	1.3	1.1	104.3	1.1	0.7	107.3	0.8	0.8	103.8	0.5	0.4	112.8
Total	7.4	7.2	106.6	8.5	7.5	108.7	6.8	5.9	102.2	5.2	4.8	109.5

a. Per 1,000 live births.
b. Number of males per 100 females.
c. Because the one-child policy creates an incentive to underreport female births, the sex ratio at birth was also calculated using the reverse survival method. Assuming that there is less of an incentive to underreport female deaths than surviving female children, this method is deemed to yield a more reliable estimate. The method estimates the number of male and female births by comparing the number of deaths of male and female children at birth with the total number of male and female children surviving at the end of the first year of life.
Source: Burgess and Zhuang 1996 (based on 1990 census data).

urban Sichuan.[21] The pro-male bias in health good expenditures is more pronounced for poorer and less diversified households in rural Sichuan; such bias remains insignificant in the split samples in urban Sichuan and rural Jiangsu. For education goods, both discrimination results detected in the overall sample (10–14 age group in Sichuan and 15–19 in Jiangsu) were more prominent in the poorer and less diversified subsample; there is no evidence of bias in the richer and more diversified subsample. For education services, the pro-male bias in post-secondary education spending detected in the overall sample for both provinces carries over to the poorer subsample. This discrimination disappears for the rich subsample in Jiangsu, but a clear pro-male bias remains in Sichuan.[22] Thus as household budget constraints increase so does discrimination against girls. Removing incentives to skew investments in secondary and tertiary education toward boys would increase the earning potential of the other half of the population.

The 1990 census data show that detected biases in household spending on health and education correspond to observed biases in age-specific mortality and educational attainment (table 3.11). Rural mortality rates and sex ratios (males per 100 females) for the first year of life are revealing. The first finding is consistent with the results reported above for health expenditures: there is no discernible gender difference in mortality rates in rural Jiangsu, but the female mortality rate is significantly higher than the male rate in rural Sichuan. However, sex ratios in the first year of life are more skewed in rural Jiangsu (115.5) than in rural Sichuan (112.6). This finding is inconsistent with both the mortality and the expenditure results and suggests that the higher sex ratio in rural Jiangsu must be due to differential treatment prior to the first year of life—providing support for the hypothesis that wealthy Jiangsu residents are more likely to abort female fetuses because they have greater access to in utero sex detection methods. After birth, there is no evidence of discrimination in health spending. The pattern in rural Sichuan is entirely different. The sex ratio at birth is moderately skewed, suggesting limited differential treatment prior to birth, but it increases significantly in the first year of life as a result of gender biases in mortality that appear to be driven partly by discrimination in health spending.

Gender gaps appear in rural education outcomes as well. Rural enrollment is markedly higher for boys than for girls across all age groups (table 3.12). The gaps are relatively small for the 6–9 age group but pronounced for the 15–19 age group. These are consistent with earlier findings of a significant pro-male bias in education

TABLE 3.12

School enrollments in Sichuan and Jiangsu, 1990

(per 100 of same sex in age group)

	Rural Sichuan		Urban Sichuan		Rural Jiangsu		Urban Jiangsu	
Age	Male	Female	Male	Female	Male	Female	Male	Female
6–9	80.1	76.6	79.8	79.8	89.6	87.1	84.7	84.6
10–14	83.1	73.2	89.3	86.9	95.3	88.9	96.9	96.1
15–19	25.7	16.1	38.2	33.9	35.0	22.0	55.4	48.4

Source: Burgess and Zhuang 1996 (based on 1990 census data).

TABLE 3.13

Off-farm employment by gender in Sichuan and Jiangsu

	Sichuan		Jiangsu	
Age and gender	Share of labor force working off-farm (percent)	Years of schooling	Share of labor force working off-farm (percent)	Years of schooling
Under 20				
Male	9.2	7.0	40.8	8.4
Female	4.7	6.3	38.3	7.2
20–29				
Male	18.8	7.6	48.6	8.7
Female	6.8	6.7	35.4	7.1
30–54				
Male	15.9	6.3	39.2	7.4
Female	2.5	3.4	16.6	3.5
Over 54				
Female	8.2	4.1	20.1	4.4
Male	1.7	0.8	3.2	0.6

Source: Burgess and Zhuang 1996.

in rural areas for this age group. As expected, enrollment gaps are less pronounced in urban areas.

The finding of gender discrimination in household health and education expenditures is important. Lower investment in girls in an environment of increasing returns to human capital portends a widening gender gap in the workplace. Data indicate that differential treatment of men and women translates into different job market outcomes (table 3.13). Lower investment in girls' education appears to restrict their access to (higher-paying) off-farm employment. Gender discrimination in intrahousehold allocation is influenced by income, however, and so is amenable to change through the growth process or targeted government policies. In fact, national data show a narrowing of the gender gap in educational attainment at all levels of schooling (table 3.14).

Do unequal outcomes in the workplace simply reflect the differential levels of educational attainment between men and women, or are other forces at work? On average, Chinese women earn between 80 and 90 percent of what men earn—much higher than the worldwide average. There is, however, evidence of occupational segregation in China; women are disproportionately represented in lower-paying jobs. Some studies also find evidence of a persistent and unexplained gap between male and female wages, even after controlling for worker characteristics, suggesting that men and women do not get equal pay for equal work (Bauer and others 1992; Meng and Miller 1995; Yang and Zax 1996). Moreover, this unexplained wage gap is high by international standards. It appears more pronounced in nonstate firms than in state enterprises. Expanding employment in nonstate firms and waning egalitarian wage policies in state firms might increase the discriminatory component of the wage gap and further erode women's relative incomes. In nonstate firms there is also an emerging bias against hiring women of

TABLE 3.14

Female school enrollments, 1980, 1990, and 1995

(percentage of total students)

Level	1980	1990	1995
Primary	44.6	46.2	47.3
Vocational middle	32.6	45.3	48.7
Regular secondary	39.6	41.9	44.8
Specialized secondary	31.5	45.4	50.3
Higher education	23.4	33.7	35.4

Source: China Statistical Yearbook 1996.

child-bearing age. Female university graduates in search of employment appear to be increasingly bypassed in favor of their male counterparts (Riley 1995).

Restructurings in state firms are resulting in layoffs, furloughs, and early retirement. Evidence suggests that women bear the brunt of all three adjustments. A 1995 survey of seven provinces and four cities found that 56 percent of laid-off workers were women. A similar survey in 1996 found that the share of women in total layoffs (60 percent) in five provinces was much higher than the ratio of women in the labor force (37–40 percent). This disparity may partly reflect the fact that sectors that are experiencing difficulty, such as textiles, employ mainly women. It also represents a rational response from the perspective of maximizing household welfare, driven by existing biases in the allocation of housing. A survey of five cities found that twice as many men as women had housing assigned by their work units. If layoffs also lead to loss of housing, it is clearly sensible for individual households as well as society at large to preserve the jobs held by men. Among women this trend is not limited to older employees, as women are increasingly being laid off or asked to retire very young. The statutory retirement age in China is 55 for women and 60 for men. Thus early retirement implies that women are leaving the labor force in their forties and sometimes even earlier. A 1996 survey of 224 enterprises in Jiangxi Province found that 53 percent of laid-off women had been asked to retire, and their retirement ages were mostly between 30 and 40. In turn, this means that retired women will have lower pensions.

International experience shows that growth alone is insufficient to eliminate gender discrimination. In China, despite considerable gains in educational attainment and rising incomes, the reforms underpinning the country's remarkable growth performance may be eroding the relative position of women. This development has profound consequences not only for women's direct contribution to economic growth but also for their position within the household and status within society. Higher levels of maternal education and a stronger voice for women in household allocation decisions have beneficial effects on the health and nutritional status of children, which are important determinants of societies' future productivity.

The case for public action to eliminate gender discrimination in access to health, education, and jobs is clear on both growth and equity grounds. It is difficult to legislate or interfere with intrahousehold resource allocation decisions, but policies can target girls' education grants, for example. Labor markets are more amenable to public action, which should aim to eliminate discrimination on the basis of gender. Policies should equalize the retirement age for men and women, remove gender biases in nonwage benefits such as housing, and eliminate wage differentials in the marketplace. These moves would also help reduce intrahousehold gender bias with respect to investing in children. The difficult question is who should pay for the costs associated with bearing and rearing children; possible answers include the mother, through lower pay or reduced employment opportunities, the parents, or society at large. These are complex decisions whose answers should reflect societal consensus. But until parenting truly becomes a partnership between men and women and women have an equal voice in formulating societal preferences, the absence of government leadership will mean that women will continue to bear the full cost of raising children while benefits accrue to all of society.

Notes

1. Background papers were prepared on various aspects of China's income inequality using micro-level data from the State Statistical Bureau's household surveys. Ravallion and Chen (1997) use a multi-year data set that allows for examination of changes in inequality and their determinants. Papers by Robin Burgess use 1990 household survey data (both rural and urban) from Jiangsu and Sichuan provinces. While changes over time cannot be inferred from a single-year survey, the inclusion of a relatively wealthy coastal province (Jiangsu) and a poor and populous interior province (Sichuan) allows for a rich set of results on the determinants of income, welfare, inequality, and gender bias. Both data sets have been adjusted to value own-grain consumption of households at market prices; Ravallion and Chen also introduce regional prices into the analysis and revalue housing and consumer durables to include the amortized flow of services rather than current cash expenses.

2. To examine the effect of ownership on income, ownership forms in the urban data set are categorized into state, collective, and privately owned work units. Workers not employed by state or collective units are categorized as employees of the private sector.

3. As an extension of the analysis, the shares of schooling and employment by ownership were included as explanatory variables. The results for 1991 and 1995 further demonstrate the strong correlation between tertiary education and labor income, and the high return to private sector jobs.

4. Another plausible explanation for increasing returns to state-owned units is the gradual monetization of in-kind income.

5. In a regression of urban income growth on initial income, a dummy variable for location in a coastal province, and the share of employment by ownership, ownership coefficients were found to be small and insignificant.

6. This section draws on Burgess (1997).

7. The analysis was done using both the Gini coefficient and log deviation (or Theil index), but because the conclusions are robust to the choice of inequality measure, only the results of the decomposition of the Gini coefficient are reported here.

8. The question of transient poverty, as distinct from chronic poverty, is an important one and has received some attention in empirical work on China (Jalan and Ravallion 1996).

9. The first year of the data set is 1988. The tendencies identified here would likely be stronger if the analysis could be carried out starting with the launching of urban reforms in 1985.

10. The analysis here is based on data provided by the State Statistical Bureau, as described in the note to table 3.5.

11. Knight and Li (1996) show that large differences in educational attainment remain even after controlling for different characteristics (sex, age, minority status) of rural and urban households.

12. Ravallion and Chen (1997) show that the share of inequality attributed to any income determinant is the product of three things: the partial regression coefficient of income on that determinant, the simple correlation coefficient with income, and the ratio of the standard deviation of that determinant relative to the standard deviation of income.

13. The remainder is made up of housing, fixed assets and financial assets (McKinley and Griffin 1993).

14. By the standard developed in the national survey, 12.7 percent of the total population and 17.4 percent of the landless population is poor.

15. Because McKinley and Griffin include the landless in their calculations, however, their Gini is higher.

16. Interestingly, and unlike other countries, the distribution of landed wealth is not a good predictor of the distribution of other assets. In fact, households that hold less land (in terms of value) tend to hold relatively more other assets.

17. There is strong evidence for this in Sichuan (Burgess 1997).

18. The same is true of fishponds, whose contribution to inequality was still low in 1990 (1.4 percent for adjusted income) but which accounted for 23 percent of the increase in inequality between 1985 and 1990.

19. Interestingly, analysis suggests that given incomplete markets, making the distribution of land less egalitarian would hamper productive efficiency, equity, and welfare, suggesting that lump-sum land redistribution in rural China may represent a rare example of a redistributive policy intervention that enhances both equity and efficiency (Burgess and Murthi 1996). Although this does not mean that the overall allocation of resources in rural China is optimal, it does suggest that the development of factor markets (particularly for labor) is likely to be more important than land redistribution in increasing returns to labor and improving household welfare.

20. These findings point to the absence of discrimination in the allocation of health services for both provinces. However, the analysis likely has limited power to pick up discrimination in this area, given the heavy subsidization of health services in 1990. Also, if attending a clinic is not costly but drugs and other health goods need to be purchased in the market, a preference may be expressed more forcibly in decisions to purchase health goods rather than in decisions to attend clinics.

21. The samples were split according to expenditures (per equivalent adult) and the share of off-farm income in net income.

22. As regards education services, the pro-male discrimination detected in the full sample for the 15–19 age group in Sichuan appears to be more pronounced for more diversified households; this might reflect the fact that it is mainly diversified households that engage in post-secondary education. Less diversified households also exhibit a pro-male bias in investments in secondary education services that was not apparent in the full sample.

chapter four

How Policies Affect Individual Welfare

People's different endowments suggest that inequality in outcomes is not only unavoidable but also that it can help nourish creativity and spur growth. As a result most societies tolerate some inequality in income. How much depends on the historical and cultural factors shaping each society's preferences. China's income inequality may continue to rise as the country's transition unfolds. But increasing inequality need not undermine growth or social harmony—so long as growth is broadly based, policy biases are eliminated, opportunities are equalized, and the poor and vulnerable are protected. This chapter shows how policies can harness growth to improve the welfare of the poor and curb damaging increases in inequality (figure 4.1).

FIGURE 4.1

Where the poor are, 1995

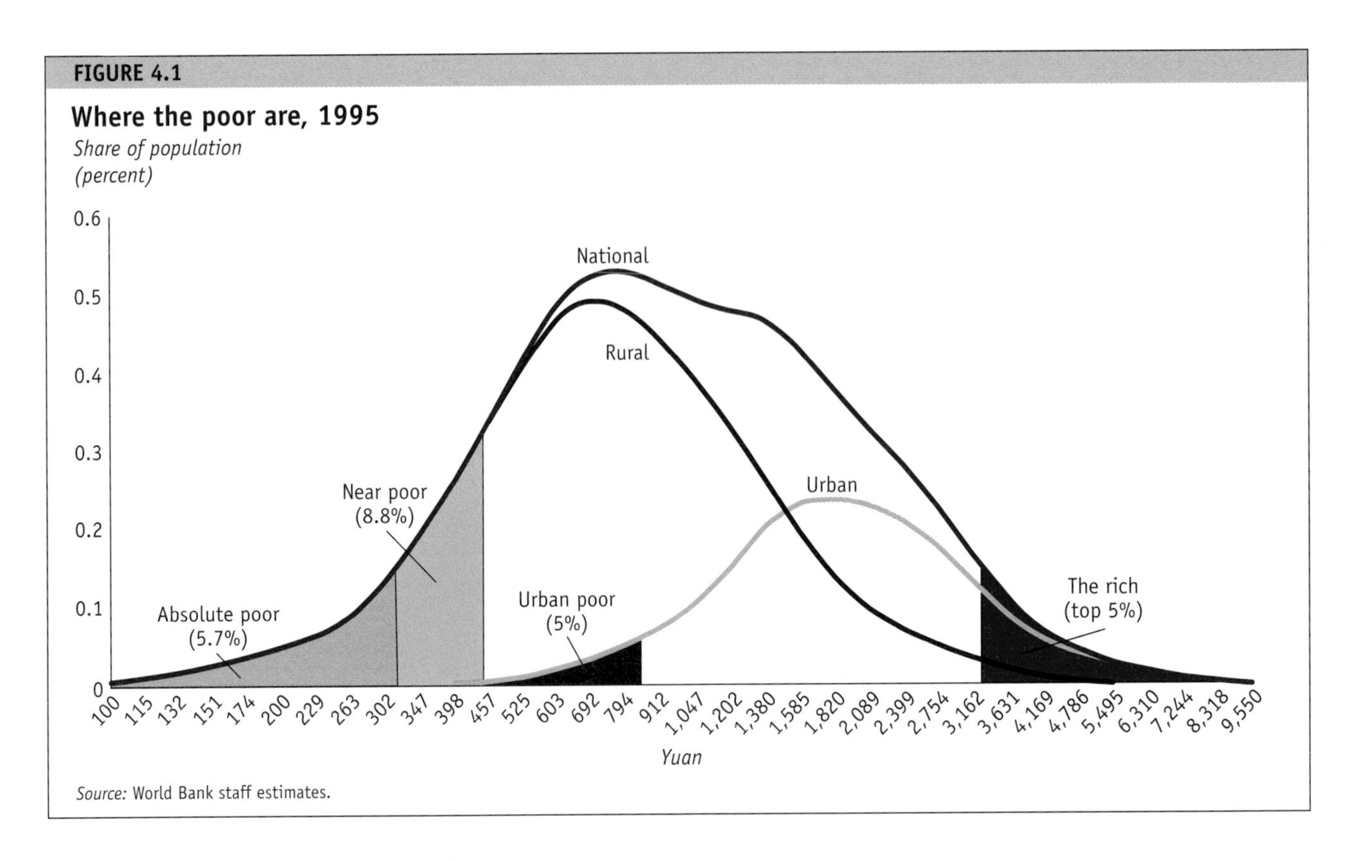

Source: World Bank staff estimates.

Eliminating policies that favor the better-off

Despite its increase over the past fifteen years, income inequality remains moderate in China. Unlike in some Latin American countries, wealth in China is not highly concentrated, and the gap between the haves and the have-nots is not large enough to threaten the social order or warrant remedial policies. The per capita income cutoff for the richest 5 percent of the population (3,180 yuan in 1990 prices) is exactly ten times the absolute poverty line (318 yuan in 1990 prices). Thus public policy should continue to focus on fostering conditions for people at the lower end of the distribution to participate in economic growth.

Some policies, however, exacerbate the gap between the rich and the poor. Eliminating policies that favor the rich would help reduce inequality. So would strengthening policies that tax the rich and that guard against unscrupulous wealth accumulation (box 4.1). Two efforts are crucial in this regard.

First, the urban bias in current policies should be redressed. Housing, food, migration, credit, state employment, and other policies provide de facto subsidies for urban residents. By preempting public resources that could be targeted to more needy populations, some of these policies directly lower the welfare of rural residents. Others do so directly. Adjusting these policies would more likely affect middle- than upper-income individuals.

Second, the coastal bias in economic policies should be removed. The natural and human capital advantages of the coastal provinces are sufficient to attract foreign investment without preferential policies. The government is already moving in this direction by eliminating tariff and tax benefits for foreign investors. The impact on regional growth patterns of policies that favor the interior is more complex. Research is needed to determine the potential effectiveness of a package of such policies. International experience with regional development efforts generally has been negative, but there has been little systematic analysis of this important issue.

There is, however, substantial evidence that well-designed intergovernmental grants can reduce public expenditure disparities across localities. The equalizing nature of China's current interprovincial transfer scheme has been eroding, and the government plans to reform it (World Bank 1995a). Income inequalities that result from the unequal size of provincial purses could be addressed by a new transfer scheme based on provincial

BOX 4.1

Dealing fairly with the rich

Almost the entire top 5 percent of China's rich (in per capita income terms) resides in urban areas (see figure 4.1). Moreover, urban incomes at the upper end of the distribution are almost certainly understated, particularly because nonwage income is not adequately captured in survey data. As a result it is difficult to draw an accurate profile of wealthy Chinese. But survey data indicate that they are much more likely to work for foreign companies or joint ventures than for the state.

Policies toward the rich should focus on bringing China's personal income tax closer to international norms and reinforcing the government's regulatory and anticorruption function. As currently administered, the personal income tax neither mobilizes resources nor redistributes income. The high threshold for exemption (800 yuan a month per type of income) implies that almost no one pays the personal income tax. The exemption level would need to be halved and combined for all sources of income to capture even 1 percent of the total population in the tax net. The role of the personal income tax also needs to be reconsidered: the administrative costs associated with the tax are not warranted unless the tax base can be expanded substantially. This goal could be accomplished by replacing the current system of schedular taxes with a comprehensive income tax (including capital gains) and by allowing increases in nominal incomes to erode the value of the current exemption threshold.

The government's current efforts to address ill-gotten wealth are appropriate. Opportunities for rent seeking must be eliminated and financial markets must be regulated. Bureaucratic discretion and access to insider information generate most illegal wealth, with negative consequences for both economic well-being and social harmony.

expenditure needs and revenue generation. Progress in reforming the system has been hampered by political concerns. Accelerating interprovincial disparities require the central government's concentrated effort if the trend is to be checked.

Protecting the absolute poor

About 70 million Chinese lived in absolute poverty in 1995, down from 270 million in 1978. Although continued growth in agriculture and off-farm employment should raise the living standards of some of the remaining poor, targeted poverty interventions will remain essential for most. Absolute poverty in China is now concentrated in remote upland areas where people eke out a living in the face of severe resource constraints (World Bank 1997b). Although these poor have land use rights, in most cases the land is of such poor quality that it is impossible to achieve subsistence crop production. Consequently, most poor people consume grain and other subsistence foods beyond their production levels and suffer when the prices of these products increase. The poorest households often are further disadvantaged by high dependency ratios, ill health, and low educational attainment. In many of China's poorest towns and villages at least half of the boys and nearly all of the girls do not attend school and will be illiterate. The poorest households have neither the physical nor the human assets to reap the benefits of growth.

Do current programs reach the poor?

The Chinese government is committed to reducing poverty, and most government ministries and agencies have special poverty reduction projects. In 1986 the State Council established the Leading Group for Poverty Reduction to coordinate poverty initiatives and to infuse rural social and relief services with a new emphasis on economic development programs in poor areas. To be eligible for development assistance, counties were designated as poor in 1986 using county-level rural income data gathered by the Ministry of Agriculture (box 4.2). The resulting list of 327 poor counties captured many of the poor. But the decision to focus on counties (rather than townships, for example) and to use county-level data to determine eligibility limited the government's ability to target more effectively. The switch to State Statistical Bureau data (county average rural per capita income) in 1992 improved the information base for targeting, but problems remain: there is substantial income variation within counties and considerable fluctuations in per capita income over time.

Analysis based on the rural household survey data for the four-province data set described in previous chapters shows that during 1985–90 roughly half the poor did not live in either the nationally or the provincially designated poor areas (table 4.1).[1] This finding suggests that there is considerable variation in per capita incomes around the county mean. Indeed, for the pooled sample (excluding Guangdong) the standard deviation of per capita consumption for the designated poor counties was close to the mean.

A single year is not a useful time period for defining eligibility for poverty alleviation schemes. Individuals may dip below the poverty line in a given year and later rise above it again. As such, they are members of the transient rather than the chronic poor. Transient poverty appears prevalent among the households sampled in 1985–90,[2] suggesting that poverty alleviation policies based on consumption (income) levels in any one year are less efficient than policies based on more extensive data.[3] In China targeting is geographic (based on county means), so household-level fluctuations in annual per capita consumption will have little impact on targeting efficiency unless there are risks that affect entire communities.

BOX 4.2

Designation of poor counties in China

In 1986, 327 "national poor counties" were designated as eligible for the central government's Poor Area Development Program on the basis of rural income and production data collected by the Ministry of Agriculture. A county was eligible if its average annual rural per capita income was less than 200 yuan in 1985 or its average per capita grain production was less than 200 kilograms a year. Political considerations also entered the calculations. If a county was an early revolutionary base, it could have a per capita rural income between 200 and 300 yuan and still qualify. Another 372 counties were designated "provincial poor counties" in 1986, eligible to benefit from province-level special assistance. Provincial poverty lines varied, ranging from 200 to 500 yuan.

In 1992 the government began using the State Statistical Bureau's household data to designate poor counties. Using these data, a county was considered a national poor county if its average annual per capita income was less than 400 yuan in 1992 and the county was not on the list of counties designated as poor in 1986, or if its average rural per capita income was less than 700 yuan in 1992 and the county was on the 1986 list. In 1996 there were 592 national poor counties.

TABLE 4.1

Do the poor live in designated poor counties?

(percentage of provincial poor living in designated poor counties)

Province	1985	1990
Guangdong	34	42
Guangxi	48	49
Guizhou	49	53
Yunnan	40	43
Total	45	48

Source: World Bank staff estimates based on State Statistical Bureau data.

Adjusting the government's approach to alleviating poverty

As poverty declines, the need for better targeting increases. Targeting to the level of townships, or perhaps even administrative villages, would reduce costs and increase the effectiveness of poverty alleviation programs. Current poverty alleviation criteria are compromising the government's ability to reach all of the poor. They also raise the cost of lifting an individual out of poverty because increasingly large numbers of people who are not poor continue to benefit from government programs.

The government should also consider refocusing its poverty reduction strategy. Most of the poor now live in remote, sparsely populated regions with low-quality land. Some will no doubt benefit from rural infrastructure investments, efforts to improve agricultural productivity, and local off-farm employment opportunities—important pillars in the government's strategy to combat poverty. In particular, those who fall into poverty because they are exposed to discrete shocks would benefit substantially from opportunities to diversify risk through off-farm employment. But returns to interventions that were effective in the past must be declining, given the changing profile of the poor. Thus a renewed emphasis on basic education and health services for the poor is essential, combined with help finding employment in economically advanced areas (box 4.3).

Evidence suggests that health emergencies contribute to transient poverty (World Bank 1996b). There is a clear need to ensure essential health services for the poor and to strengthen public health programs. Chronic poverty in China is highly correlated with poor health, low levels of educational attainment, and illiteracy. Poor households should be compensated for the direct cost—and possibly some of the indirect (opportunity) costs—of educating their children. To increase returns to schooling, labor mobility should be promoted and the quality of education improved. Doing so

BOX 4.3

Helping workers relocate: A strategy for alleviating poverty

For many workers, particularly members of the surplus labor force, securing off-farm employment is the only way to shake off poverty. Yet these jobs are often out of reach. Although rural-urban migration exploded in the late 1980s and early 1990s, the absolute poor have not benefited much because the remote villages where they live lay outside the information network that is so critical to fostering migration. Most of the poor have no way to learn about jobs and no money for transportation to job sites. And if they do find jobs on their own, their low level of education makes them vulnerable to exploitation. Yet factory managers in both the coastal provinces and more developed interior provinces are willing to hire workers with minimal education.

The World Bank–assisted Southwest China Poverty Reduction Project, under implementation since 1995, is a multisectoral attack on poverty in the worst-affected unplanned areas of Guangxi, Guizhou, and Yunnan. Part of the project aims to provide safe, secure off-farm jobs for rural laborers. Working through provincial and county labor offices, the project identifies job opportunities. Participants receive training in job skills and safety issues. They can take out loans under the project to cover transportation costs, training, and supplies. The project monitors the work site to ensure that the jobs are safe and the workers are well treated.

The project is demonstrating that organized labor mobility is a powerful and cost-effective tool for alleviating poverty. By the end of 1996 the project had helped about 50,000 rural poor secure off-farm jobs. Workers are earning 400–800 yuan a month, and most are able to send money home after a few months of employment. These remittances, averaging around 150 yuan a month, easily lift families in the project area over the poverty line. Families use remittances to buy grain and farm inputs and to fund schooling.

Source: World Bank staff.

would reduce parents' reluctance to send their children to school. Information from household surveys and special surveys of migrants shows the importance of remittances to incomes, including those of the poor. The main constraint to increased migration appears to be access to information about job opportunities. A recent village-level study on migration flows found that familiarity with previous migrants was the most important determinant of migration (Rozelle and others 1997). Expanding the government's already important activities in this area offers promise.

Boosting the potential of the near poor

China's "near poor" have incomes above the absolute poverty threshold but are considered poor by most international standards. By this report's definition, the number of people considered near poor dropped from some 200 million in 1981 (25 percent of the population) to 100 million in 1995 (8.8 percent of the population).[4] This segment of the population has benefited greatly from reforms in grain pricing and strong growth in off-farm employment.

Policies that are beneficial to the absolute poor will also help improve the welfare of the near poor. Increased availability of basic education and health services is essential to ensure that the poor are not left out of the growth process. Access to higher levels of education, a more accommodating grain policy, greater integration in labor markets, and better-functioning credit markets would further upgrade the living standards of the near poor.

Educating the near poor

Education is an increasingly important determinant of income. Arresting deterioration in China's income distribution will require ensuring greater equality in access to high school education and above. To this end, the government should consider providing merit-based assistance to poor families to help defray the increasing costs of higher education.

Developing a more flexible grain policy

Government grain production policies continue to depress rural incomes. In rural China low incomes are closely associated with grain production (figure 4.2). Higher returns accrue to nongrain agriculture and to off-farm employment in particular. Survey data from the four southern provinces show that among the near poor grain income accounted for nearly half of per capita income in 1990. Although grain incomes have increased substantially, thanks to higher yields and price increases, the potential for further gains in this area is limited. In particular, market prices in China's cities are now close to—and in some cases even higher than—world market

FIGURE 4.2

Dependence on grain income declines as per capita income increases, 1990

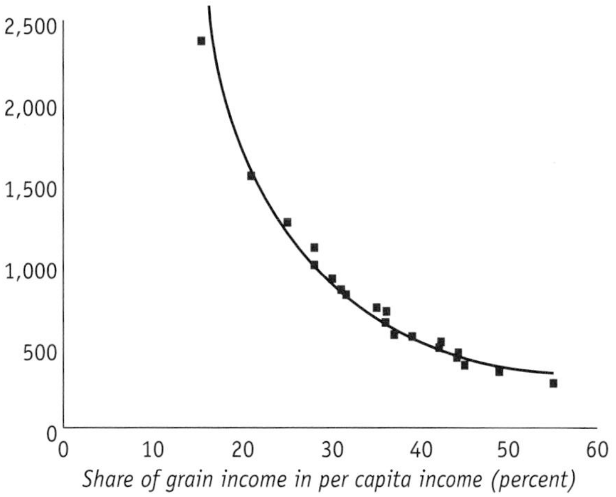

Mean per capita income (yuan)

2,500
2,000
1,500
1,000
500
0

0 10 20 30 40 50 60

Share of grain income in per capita income (percent)

Note: Data are for 5 percentile groupings ranked by per capita income for Guangdong, Guangxi, Guizhou, and Yunnan. Incomes reflect the Ravallion and Chen (1997) adjustments (see chapters 1–3).
Source: World Bank staff estimates.

prices for most grains. Grain farmers would benefit from the alignment of procurement prices with (hitherto higher) market prices, more efficient transport and distribution, more effective fertilizer use, and improved agricultural research and extension.[5] But the biggest benefit would come from reallocating labor within farming and between farming and nonfarming.

The government's desire to ensure that China is 95 percent self-sufficient in grain erodes the welfare of grain producers. This policy has become more varied and less intrusive over time but it continues to restrict farmers' choices. The policy of "protecting arable land" locks certain areas into grain (and cotton) cultivation. In addition, farmers must meet their quota obligations to the state. And since 1995 the governor's responsibility system for grain self-sufficiency has pushed responsibility to each subsequent administrative level and down to the village (box 4.4). At the same time, authority to import grain, previously at the discretion of the coastal provinces, was centralized in Beijing. Such constraints appear to stack the deck so that farmers have little choice except to grow grain. As a result excess labor is used to cultivate grain in areas where returns are generally low. Relaxing such policies would benefit farmers by allowing them to engage in higher value-added activities, or on off the farm.

BOX 4.4

The "rice sack" system

In 1995 the Chinese government launched the "rice sack" governor's responsibility system. The system requires provinces to balance the supply and demand of grain and maintain the stability of the grain market. It was put in place in response to slow growth in grain production in the early 1990s: between 1990 and 1993 the average annual increase in grain production was less than 1 percent. Stabilizing production in the south and east is a prime focus of the policy.

Provincial leaders, feeling a greater sense of responsibility for grain production, have taken steps to ensure that a specified number of acres is sown with grain, to boost yields per unit area, and to increase total grain supplies. To boost grain output, many provinces have increased funding for agriculture and agriculture technology. Investments in irrigation have also risen. To motivate farmers to grow grain, twenty provinces now provide extraprocurement price subsidies. These subsidies ranges from 2 to 30 yuan for every 50 kilograms of grain produced. In addition, fertilizer purchase is now linked to grain production in many areas. For every 50 kilograms of grain farmers sell to the state, they are given 5 to 10 kilograms of fertilizers. The system is credited with boosting grain production in coastal areas. Grain output slipped in Guangdong, Fujian, Zhejiang, and Jiangsu between 1990 and 1993, but increased between 1993 and 1995.

Administrative measures to boost grain production may provide short-term palliatives, but they obstruct progress in addressing the underlying issues affecting grain production in China, which have to do with improving the distribution system and investing in agricultural technology. The system's main problem is that farmers simply do not have sufficient incentive to grow grain. Especially in southern and eastern provinces, off-farm labor opportunities are abundant and the opportunity cost for growing grain is too great.

Source: FBIS 1996a.

The obvious alternative to grain self-sufficiency is increased reliance on food imports. Attention to China's grain policies has generally come from foreigners who are concerned about the missed opportunities for exports to China or about the implications for world food prices of a more import-reliant China. The effects these policies have on the welfare of China's own farmers has received much less attention. China's concerns about relying on potentially volatile foreign grain supplies are valid, and would have to be addressed were policies to be adjusted.

BOX 4.5

Making a market for microfinance

China's rural poor lack access to credit. Most rural credit is provided through state agencies to rural enterprises and state-owned collectives, with a small portion going to better-off families. Informal credit is increasing but is expensive and tends to miss poor areas. As a result households with potentially profitable economic activities may remain in poverty for lack of credit.

The Chinese government recognizes the potential of microfinance in reducing poverty and has supported several microfinance pilots in cooperation with international donors. If these experiments prove successful, the government plans to expand microfinance and make it an integral part of its antipoverty efforts.

Although microfinance can be highly effective in alleviating poverty, the potential damage from a badly structured program can be great. If the wrong households are targeted, programs become financially unsustainable and drain government resources. And when microfinance institutions fail, the experience can easily sour an entire region on the potential benefits of microfinance.

China's pilot programs need to be adjusted if the microfinance experiment is to succeed. Programs currently target grain-deficit families living in areas far from rural markets. Placing such families in debt harms the households and endangers the program's sustainability. The next tier of (near-poor) households can benefit much more from microfinance and should be the main targets. Also, credit products should be better designed, linking repayment schedules with households' main economic activities.

The financial structure of the pilots also raises major obstacles to sustainability. Few programs maintain accounting records, so they are unable to track the costs of their services. Interest rates and fees do not cover the high costs of serving low-income, dispersed rural populations. And because microfinance entities are not registered as financial institutions, they cannot collect savings. Ensuring that future reforms of the financial sector create a friendlier environment for microfinance is key to the long-term success of the Chinese experiment.

Serving the creditworthy

Increased diversification into off-farm employment will not only help increase the incomes of the near poor but also will reduce their vulnerability to income fluctuations. More education is key for improved market access, but better rural infrastructure and credit availability will also help bring markets within reach and get new businesses started. Microfinance, a relatively new instrument in China, has considerable potential to boost rural incomes if it is properly structured (box 4.5). International experience shows that microfinance programs can contribute to poverty reduction and be sustainable, but only when families are able to invest in productive activities that can generate cash flow for repayment. Thus such programs are likely to be well suited to the near poor but will fail if used to aid the destitute.

Integrating labor markets

The benefits of increased labor market integration for the near poor are clear. Despite considerable relaxation in the rules and regulations governing rural-urban migration, forms of control and other impediments remain. Controls in both sender and recipient regions reflect the government's intolerance for "blind" migration—that is, migration into cities without a job. Most migrants to China's major cities are forced out by three no's—no *hukou*, no housing, and no job. In addition, the uncertain status of migrants in urban areas, the absence of a housing market, and the unavailability of social services dampen the demand for migration (annex 1). The authorities' desire to control the pace of rural-urban migration is understandable. The potential costs in terms of greater urban congestion, higher incidence of crime, and dislocation of established urban workers may be sizable. But it is important to weigh these costs against the considerable benefits of labor mobility, not just for the individual migrants but also for the economies of the host cities and the families left behind. The government could help broaden access to migration opportunities by strengthening job information networks.

Caring for the urban poor

The near-absence of poverty in urban China is unusual. In most countries urban poverty contributes significantly to overall poverty (figure 4.3). In China, by contrast, available data indicated that no urban residents have incomes below the absolute poverty line. And in 1995 just 0.1 percent of the registered urban population lived below the higher poverty threshold, down from a peak of 1.8 percent in 1989. Low urban poverty reflects China's limited urbanization and the continued segmentation of its urban and rural economies. Not only is urban poverty negligible, but urban inequality is also low by international standards. It is likely that both will rise in the future. But if this increase is due to a decline in the rural-urban gap as a result of greater labor mobility and increased market integration, it should not increase overall inequality.

Who are the urban poor?

The characteristics of the poorest 5 percent of urban households (ranked by per capita income) can be compared with those of the average urban resident to yield information about the relative urban poor. As table 4.2 shows, poor urban households have more members and fewer income earners. The drop in female employment is particularly notable. Members of poor urban households are more likely to work in collectives, as domestics, or be self-employed. But above all, they are more likely to be unemployed: only 1.8 percent of the members of the average urban household are waiting for a job or a job assignment; in poor households the corresponding figure is 6.3 percent. Poor households are underrepresented in knowledge-intensive occupations and are far less likely to work for the Communist party and the government. They are also less likely to be employed by the state or by joint ventures and are less likely to have members who are retired. Unlike in other developing and transition economies, retirement in China does not push people into poverty, reflecting the still generous pensions received by retirees.

Information from a survey of five cities (Beijing, Shanghai, Chongqing, Guangzhou, Shenyang) complements these findings and provides additional insight. The survey found that retirement does not increase the likelihood of a household falling into poverty, but unemployment and furloughs do.[6] Survey results also show a much higher incidence of unemployment than is indicated in State Statistical Bureau data—and one that is rising. Even aside from those who took early retirement, adding together reported unemployment and those who were furloughed brings total unemployment to 3.0 percent in 1991 and 8.2 percent in 1995.

Early retirement and furloughs are being used to shed excess workers in enterprises, according to the survey. Among those who had retired since 1991, 10 percent were in their forties or younger and 38 percent were in their fifties. The statutory retirement age

FIGURE 4.3

Unlike in most countries, urban poverty in China is negligible

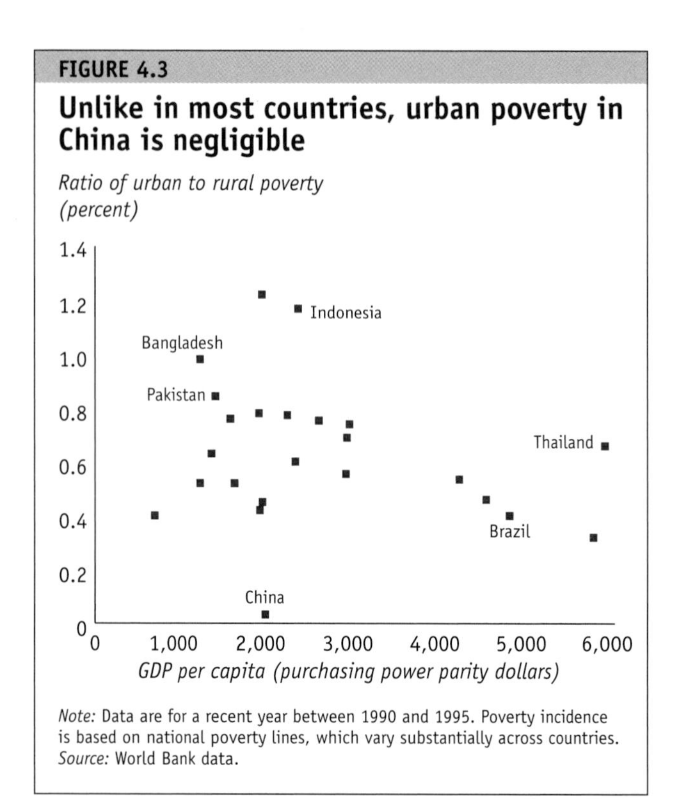

Note: Data are for a recent year between 1990 and 1995. Poverty incidence is based on national poverty lines, which vary substantially across countries.
Source: World Bank data.

in China is 55 for women and 60 for men. About 12 percent of the individuals in the sample changed job status between 1991 and 1995. The bulk of these changes came from the state sector (60 percent), and they were almost entirely for the worse. Among state employees who changed jobs, less than 30 percent found work in other sectors; the rest retired (45 percent), left the labor force (10 percent), became unemployed (9 percent), or were furloughed (8 percent). New entrants into the labor force went primarily into private and foreign jobs, reflecting the locus of new job creation and the continuing decline of state and collective enterprises.

Dealing with rising urban poverty

China's cities have been changing rapidly and will continue to do so for at least the next two decades. China remains underurbanized. This will change, and as it does there will be a substantial transformation of both the structures and the faces that make up modern China. Poverty and unemployment have been almost unknown to China's urban residents, but this too is changing. As wages increasingly reflect productivity differentials and production structures adjust to China's comparative advantage in international mar-

TABLE 4.2

Characteristics of the urban poor, by household income

Characteristic	Number per 100 households		Share of household members (percent)	
	Average income	Bottom 5 percent	Average income	Bottom 5 percent
Household members	321.7	383.1	100.0	100.0
Income earners	234.3	209.9	72.8	54.8
Male	118.6	113.9	36.9	29.7
Female	115.7	96.0	36.0	25.1
Employed	185.6	167.7	57.7	43.8
Male	96.6	93.1	30.0	24.3
Female	89.0	74.7	27.7	19.5
Average age of household members	35.0	33.0		
Employer/employment status				
State	145.9	109.8	45.4	28.7
Collective	26.9	43.8	8.4	11.4
Joint venture or foreign-owned	2.5	0.7	0.8	0.2
Privately owned, self-employed	2.8	8.9	0.9	2.3
Privately owned, employed	1.1	1.7	0.4	0.4
Retirees, reemployed	5.5	1.5	1.7	0.4
Other employment	0.8	1.5	0.2	0.4
Retirees	35.1	25.0	10.9	6.5
Disabled workers	0.7	2.2	0.2	0.6
Household workers	4.3	21.7	1.3	5.7
Waiting for jobs	5.6	21.7	1.7	5.7
Waiting for job assignments	0.4	2.2	0.1	0.6
Students	62.5	86.7	19.4	22.6
Waiting for entry into higher education	0.1	0.1	0.0	0.0
Other	27.4	55.8	8.5	14.6
Sector of employment	185.6	167.7	100.0	100.0
Rural	2.5	3.2	1.3	1.9
Industry	69.1	69.1	37.2	41.2
Geological exploration	1.4	2.3	0.8	1.4
Construction, transportation	5.7	6.7	3.1	4.0
Post and telecom, commerce, trade	10.8	7.4	5.8	4.4
Material supply, housing, public utility	27.3	38.3	14.7	22.8
Management, resident service	8.2	7.1	4.4	4.2
Health, sports, social welfare	8.9	4.7	4.8	2.8
Culture, arts, education	14.6	6.5	7.9	3.9
Science, research, technology	3.5	0.5	1.9	0.3
Finance and insurance	4.5	0.9	2.4	0.5
Party and government, mass organization	26.7	17.7	14.4	10.6
Other industry	2.7	3.5	1.4	2.1

Source: State Statistical Bureau urban household survey team.

kets, there will be winners and losers. Thus the government must put in place a safety net for the potentially vulnerable population in urban areas. Analysis points in particular to laid-off and furloughed workers. The disabled also remain vulnerable, while women appear to suffer disproportionately from enterprise restructuring. In addition, the growing migrant population is potentially at risk. While urban migrants typically are not among the ranks of the urban poor, they are deprived of social assistance, suffer from poor living conditions and a difficult work environment, and are vulnerable to emotional distress (see annex 1).

The government needs better information to develop programs to assist the urban poor. Establishing a meaningful urban poverty line would help, as would systematic monitoring of the unemployed and their adjustment experience. It is also time for the government to take a comprehensive look at its system of social protection. Substantial work has already gone into analyzing the pension and health care finance systems. Efforts should now focus on China's system of unemployment compensation, disability and other benefits (including maternity leave), and labor training and retraining schemes. The government should also strengthen job information networks. Finally, lessons should be emerging from urban job creation and reemployment programs. Much could be learned from a systematic evaluation of these programs' cost-effectiveness and the conditions for their success or failure.

Notes

1. Using the definition in Chen and Ravallion (1996).

2. Roughly half of the mean squared poverty gap (defined as the income needed to bring the poor out of poverty) for the four provinces resulted from fluctuations in consumption, while about 40 percent of the transient poverty was found among households who are not poor on average (Jalan and Ravallion 1996b).

3. Using a current cross-section of consumptions, Jalan and Ravallion (1996b) find that the full cost of eliminating chronic poverty would be three or four times the poverty gap based on mean consumption over six years.

4. The poverty line used to identify the near poor is set at 454 yuan in 1990 prices, equivalent to $1 a day of income in 1985 purchasing power parity dollars using the Penn World Tables (Summers and Heston 1995).

5. For a detailed discussion, see World Bank (1997a).

6. Less than 6 percent of those who had retired between 1991 and 1995 became "new poor," defined as falling into the bottom two income deciles from at least one step above. Among sampled households, 27 percent of the "new poor" came from those who were newly unemployed, furloughed, or had left the labor force.

annex one

Migration and Inequality in China

Over the past decade greater employment opportunities and higher living standards have lured China's peasants to the cities in unprecedented numbers. This migration has redistributed income and in some areas appears to have increased inequality, but the exact effects cannot be measured with available data. National urban household surveys include only official urban residents; most migrants who reside in cities are not officially registered.

Although migration creates new challenges for China's rural and urban economies, so far its effects have been largely positive. Migration alleviates the pronounced inequality between poor rural people and wealthy urbanites and helps redistribute rural incomes because migrants send significant portions of their earnings back to their families. Although migration may be increasing inequality among urban residents, national inequality would likely be more severe in the absence of migration. Governments could increase the income distribution benefits of greater labor mobility by providing poor rural residents with more opportunities to migrate.

Evolution of migration and current trends

Because population policies have discouraged rural residents from moving to cities, about 70 percent of China's population is rural—unusual for a country at its level of development (China State Statistical Bureau 1996). In 1948 China enacted the household registration system (*hukou*), designating households as rural or urban. This system was iron-clad, and converting from a rural to an urban hukou was nearly impossible. Members of urban households could live in cities and small towns, received state-subsidized grain supplies, and could work in government enterprises. These rights were denied to peasants with rural hukous.

In 1980 policymakers introduced the household responsibility system, which allowed households to determine how to allocate labor between farm and off-farm activities. Although rural residents could leave the land, they still could not legally reside in cities. Without urban registration, migrants are essentially second-class citizens, and their stay in the city is subject to the whim of the authorities. And because migrants are denied access to urban services, good jobs, and social status, they are discouraged from bringing their families. Moreover, migrants tend to retain ties to agriculture because it provides security, especially valued given the lack of a formal old-age security system in rural China. So, despite the high income potential of migration, these drawbacks limit the duration of migration, and migrants return frequently to their home villages. Although urban migrants earn nearly three times as much as rural nonfarm workers, better-educated rural residents prefer local nonfarm employment.

In the early 1980s rural incomes grew rapidly, so the difficulties in finding urban employment and the impediments of the hukou system outweighed the incentives to migrate. But later, as the early gains from the rural reforms were exhausted, urban reforms boosted urban growth rates, and rising disparities spurred migration. Recent estimates of the current number of migrants range from 30 million to 200 million. Data from a recent survey suggest that there are around 40 million long-term migrants. That number swells to 70 million when it includes people who work close enough to home to return each night and self-employed workers who travel between their home village and outside locations (Rozelle and others 1997). Migrants tend to come from rural areas in the interior provinces and go to cities in the richer coastal provinces; there is also substantial migration within coastal areas. Emigration rates for poorer interior provinces (such as Yunnan) are lower.

The large influx of migrants to the cities has sparked anxiety among urban residents, and officials have scrambled to enact policies to curb and regulate the flow. These anxieties reflect both capacity concerns (an overflow of the railway system during the Chinese New Year) and understandable fears (that migrants will increase urban unemployment and crime, crowd urban schools, and evade birth control regulations). To try to stem "blind" migration, recent national policies require migrants to obtain permits from the authorities in both the source village and the destination area testifying that they are migrating to a job. Requirements for other permits have proliferated and vary between localities. But while the new policies create hardship for migrants, requiring payment for a patchwork of permits, they have done little to stem the tide. The authorities recognize some of the benefits of migration, and policy documents (including the Ninth Five-Year Plan) call for orderly migration to help alleviate poverty. Recent efforts have focused on coordination between provinces to regulate the flow of migrants and mitigate congestion of the transport system.

Whether large-scale migration continues will depend in part on whether the surplus labor force has been exhausted. Chinese researchers indicate that the surplus labor force (not including those who have already migrated) ranges from 130 million to 168 million (Mason 1997a). If these figures are accurate, migration could continue for the next decade given current agricultural and population policies and urban growth rates. Survey work has shown that the shadow wage rate for agriculture is extremely low—well below that for other rural employment—supporting claims that a large surplus labor force still exists (Hare and Zhao 1996).

Migrant profile

Migration tends to be concentrated by area and appears to vary widely across counties and even villages. Most migrants find their jobs through family and other informal connections, and these connections are mostly at the village level (China Ministry of Agriculture 1996). Only 15 percent of migrants find jobs through formal channels, such as local labor bureaus and employment offices. Most migrants are young, unmarried, and work in blue-collar and service jobs. The average migrant is less educated than the general population but more educated than the rural population. Few migrants come from the ranks of the absolute poor, who lack even the few years of schooling and basic Mandarin required for most migrant jobs.

Young, single workers migrate because job opportunities are abundant and their migration costs are low. The construction sector absorbs the largest share of migrant workers, followed by manufacturing, light assembly, and services. The predominance of construction jobs is one reason men migrate more often than women. In areas where light assembly jobs dominate, however, female workers may outnumber males by as much as seven to one. Joint ventures in Guangdong Province that offer high-paying, relatively secure jobs prefer to hire women. This, and the increasing information available to women, may account for recent increases in the proportion of female migrants (Rozelle and others 1997).

In other countries educated people are the most likely to migrate, but in China migrants tend to be neither the best nor the worst educated in their home villages. In the late 1980s most migrants had only an elementary school education, but by the mid-1990s most had completed middle school. Well-educated workers tend to work in township and village enterprises or to cultivate nonstaple crops, both of which may yield even higher incomes than migration (Hare and Zhao 1996).

Migrants tend to be poorer than the average rural resident but, as noted, they generally are not from the ranks of the absolute poor. Data show that in 1994 the highest concentration of emigration from eighteen provinces in Sichuan came from counties with incomes around the average for rural Sichuan and slightly below the national poverty line. In 1995 a Ministry of Agriculture study surveyed four counties with high migration rates in Anhui and Sichuan provinces. The three that were officially designated "poor counties" had the highest migration. Migrants were typically less advantaged than nonmigrants on almost all counts: their pre-migration income levels were up to 30 percent lower; they had less cultivated land and fewer fixed assets; more of their income came from agriculture, and more from staple crops; and fewer of their household members worked in township and village enterprises (China Ministry of Agriculture 1996b).

Migration patterns are changing. Migrants now travel farther to find work and stay away for longer periods, although women tend to migrate shorter distances than do men. Recent government policy has tried to ease the pressure on the transport system by preventing migrants from returning home at Spring Festival, a custom that prevailed in the past. Shorter migration cycles are common for migrants working closer to home, who may take off-farm jobs during the agricultural slack season and return home for the harvest or at other times when they are needed. Migrants typically return home after fewer than 200 days, and some researchers have found evidence that workers migrate for five to seven years, then stop (Rozelle and others 1997).

Migration's impact on urban areas

Unlike most countries, China's development has proceeded without pronounced urban overcrowding and degradation, thanks in part to policies that have restricted migration. As a result Chinese citizens consider urban areas to be more orderly and free of crime than rural areas, so any increase in urban crime is viewed with alarm—often out of proportion to the threat it presents. Thus urban residents blame migrants for deteriorating living standards, mounting crime rates, and increasing unemployment.

Costs and benefits

Certainly, migration has reduced the standard of living in cities. The transient population commits more than 30 percent of the crime in Beijing, 70 percent in Shanghai, and 80 percent in Guangdong (FBIS 1996b). Migration may contribute indirectly to urban unemployment because migrants compete for low-wage, low-status jobs, although most urban residents face little risk of being displaced. And because migration swells the urban population, transport and services are increasingly overcrowded and strained, and since many migrants escape the tax net, they do not contribute toward maintaining municipal infrastructure.

Nevertheless, cities with large migrant populations tend to thrive. Migrants contribute to a city's vitality by providing a low-wage workforce. And by increasing the variety of products available to local residents, migrant-owned businesses stimulate the urban economy (Wu and Li 1996). For example, in 1990 Dongguan, in Guangdong Province, contained only 50 enterprises and 5,000 residents. It is now home to 20,000 foreign-invested enterprises and has a population of 500,000, of which more than 90 percent are migrants. Dongguan's economy has grown by 20 percent a year since 1990, well above even Guangdong's high growth rates.

Better off but far from equal

Migrant workers earn salaries that are many times greater than farming incomes in their home villages, and migration reduces rural-urban income disparities through remittances. But resident workers earn more money than migrants and have better employment benefits. Furthermore, most migrants' access to services does not improve and may even worsen when they relocate to the city, so the effect of migration on equality is difficult to calculate.

Migrant workers' consumption and savings are higher than those of their rural counterparts. But migrant workers earn, consume, and save less than urban resident workers. In 1995 the average wage of migrant workers in Shanghai was 704 yuan a month, about four times the average rural income in source provinces that supply the Shanghai market. Migrant workers consume a smaller share of their earnings (51 percent for migrants compared with 64 percent for residents) and save a larger share than their Shanghai counterparts. But in absolute terms, migrants' consumption and savings are lower. Migrant workers from Anhui and Sichuan provinces earn 1.64 yuan and 1.72 yuan an hour, compared with the national average wage of 2.23 yuan an hour (China Ministry of Agriculture 1996a and 1996b). Despite earning less than their urban counterparts, migrants generally do not fall into the ranks of the urban poor.

The difference in income between urban and migrant workers is due mainly to different occupational profiles. Migrant workers tend to fill low-wage jobs. More than 90 percent of the surveyed workers from Sichuan and Anhui are employed in nonprofessional positions. In Shanghai more than 60 percent of migrant workers are employed in factories or in services in nonprofessional positions. Moreover, within a given occupation migrant workers tend to be paid less than permanent urban workers. For example, a migrant textile worker in Shanghai makes an average of 500 yuan a month, while an officially registered urban textile worker in Shanghai makes 600 yuan a month (China Ministry of Agriculture 1996a). Wage gaps between migrant and resident workers would be even wider except that 10 percent of Shanghai's migrant workers have technical and white-collar managerial positions that pay more than 900 yuan a month.

State enterprises employ 65 percent of resident urban workers, and 85 percent of their remuneration comprises employment benefits such as housing, medical care, education and daycare for children, maternity benefits, vacation leave, pensions, food subsidies, and compensation for job-related injuries. Few migrant workers get jobs in state enterprises. For example, only 14 percent of migrant workers from Anhui and Sichuan work for the state. As a result disparities and inequality between migrants and residents are even more pronounced than the data on wages suggest.

Employee benefits for migrants are both meager and often of low quality. Employer-provided housing for single migrant workers usually consists of dormitories or makeshift arrangements in the workplace. Restaurant workers may sleep in the back room, construction workers live in tent structures with no amenities, and small traders often congregate in shacks and shanty towns on the outskirts of major cities.

Health and education benefits for migrants are also skeletal, and migrants are likely to incur high health expenses for basic care because, while some companies pay for treatment of job-related injuries, the compensation for migrants is far below that for regular workers. Employers pay for neither routine care nor serious illnesses—a serious problem since migrants are much more likely to live in crowded, unsanitary conditions that pose higher risks for communicable disease such as tuberculosis.

Children of migrants are not permitted to enroll in public schools or are forced to pay much higher fees than urban residents. National policy mandates that localities provide facilities for enrollment of all school-age children. But this regulation is often interpreted as applying only to permanent residents. To enroll, migrant families often must pay fees that are up to ten times higher than those for children of urban residents.

To address the lack of access to services, migrants in some cities have grouped together into makeshift villages, formed according to province or county of origin. Villages provide migrants with a wide range of services, from shops to hospitals and schools. In Shenzhen 10,000 migrant children attend hut schools, minimal facilities that fall below mandated national norms. At one point local authorities closed the hut schools because they failed inspection tests.

Both income disparities and unequal access to services are more pronounced when families migrate. Single migrants can make do without full access to health care, education, and housing, but for families the impact can be devastating. And because migrant families tend to be large, income disparities between urban residents and migrants increase with family migration (FBIS 1996b). Family migration is low in China—Ministry of Agriculture data show that only 6.6 percent of migrants bring families and less than 1 percent succeed in changing to an urban hukou. But even a small percentage can have a large effect on urban areas. For example, in Shanghai it is estimated that 100,000 of the city's 2.6 million migrants have moved their families into the city, increasing the population by more than 320,000. Municipal officials say that nearly all these families are poor (*South China Morning Post,* 5 May 1997).

Of course, in many ways the services available to migrants are no worse than what is available in rural areas, particularly for migrants from very poor communities. There is no rural pension scheme in China. Most rural Chinese now pay directly for health services, and catastrophic illnesses can easily impoverish even middle-income rural residents. And rich migrants can afford more sophisticated care than is available in rural areas. Migrants face much higher education costs than in their home village, but if they can afford them the quality is higher in cities.

Migration's impact on rural areas

The benefits to rural households from migration are dramatic, particularly for poor households. All surveys of migrants have reported high levels of remittances. Migrants typically send home 20–50 percent of their income. A migrant from Sichuan earning 5,000 yuan a year typically sends home almost 2,000 yuan, more than twice the 1995 rural per capita income in Sichuan. In Sichuan and Anhui migrant incomes account for an average 20 percent of household income and 50 percent of household cash income. Households in poor regions have a larger share of migrant income in total income than do households in wealthy regions.

Migration clearly reduces income disparities and inequality between rural and urban households. Not only does it provide more current and future income for rural households, it also reduces disparities in access to services such as health and education by providing resources to pay for such expenditures. In Anhui and Sichuan remittances are used to construct homes, meet daily living expenses, buy agricultural inputs such as fertilizer, and accrue savings. About 90 percent of the villagers surveyed who had built new homes in 1994 were migrant families. Data on migration of the absolute poor show that these families spend most remitted funds on productive inputs, daily necessities, livestock, education, and housing. In the poorest households migration income brings families out of a grain deficit.

By raising the incomes of the rural poor, migration is reducing income disparities between the richest and poorest rural areas. Within these areas, however, migration appears to increase income disparities, especially in poor regions. More people migrate out of poor areas, but the absolute poor tend to stay behind (China Ministry of Agriculture 1996b).

When families migrate, the beneficial effect of migration on urban-rural income disparities is diluted. Migrants send back much less in remittances to family members remaining in rural areas once they are joined by their immediate family.

The effect migration has on rural communities and the rural economy (beyond the household level) is unclear. The migration of young, well-educated males could easily cause agricultural productivity and grain production to suffer. But many surveys have found a positive correlation between grain production (in both absolute and per capita terms) and migration. Of course, as surplus labor leaves rural areas, productivity would be expected to increase. But it also seems that agricultural experience contributes more to productivity than does education (after completion of three years of elementary school). With higher migration rates for men, women's agricultural work has tended to increase, albeit with large regional variations. The effect of this phenomenon on productivity has not been widely studied. Many observers have noted that men who are needed for farm work tend to migrate short distances and for jobs that allow them to return in time for the harvest.

Migration appears to bring other benefits to the rural economy, including increased diversification and growth in off-farm employment and self-employment. Migrant families tend to engage in more nonagricultural activities, and counties with a long tradition of migration appear to have more nonagricultural industries. Village leaders say that rising migration has stimulated the expansion of the self-employed sector, because remittances increase the demand for services. But within agriculture, nonmigrant households invest more, earn more, and engage in a more diverse range of activities. Whether migrants return to rural areas will, of course, determine the long-term effect of migration on the rural economy. While more than 80 percent of surveyed migrants say that they plan to return, data on returnees are sparse (China Ministry of Agriculture 1996b; Rozelle and others 1997).

Conclusion

Migration has reduced rural-urban income disparities at both the individual and household levels. Migration also has lowered inequality (access to services), except in cases where families migrate. The effect on disparities within rural areas is less clear-cut. Migration appears to reduce income disparities within rural areas and lower inequality between rich and poor regions. But within regions, and particularly within poor regions, migration may increase income disparities and inequality. Migration has increased income disparities and inequality within urban areas, and the effect is much more pronounced when families migrate.

Migration has had positive effects on both rural and urban economies. Evidence does not indicate that migration has a negative effect on agricultural productivity, and it is stimulating the rural economy in many ways. While migration's effect on urban infrastructure, congestion, crime, and other problems is clearly negative, in many ways these negative effects are outweighed by the positive contributions migrants make to urban economies.

Certainly, Chinese policymakers could take steps to maximize the benefits of migration while helping groups and localities adjust to the negative effects. The poverty alleviation impact of migration has been strong, but modifying current policies could heighten this impact. For example, it appears that information networks are a strong determinant of migration and that they often miss the poorest areas. The Chinese government could increase opportunities for a broader range of citizens to migrate by strengthening these networks. County labor bureaus are a natural source of information about migration opportunities, but the extent to which they supply this information varies both across and within provinces.

Strengthening information networks would increase migration opportunities for most of China's near poor. But the absolute poor face greater obstacles than information. Lack of a basic education and a weak command of Mandarin (for members of minority groups) make it more difficult for this group to migrate and make them more vulnerable to abuse. Nevertheless, migration is a powerful means for helping the absolute poor. And because this group tends to live in China's most remote, resource-poor areas, other means of improving their livelihood are often unavailable. Experiments by the government and the World Bank to help the poorest migrate safely are promising and should be replicated.

Carrying through with urban reforms would both lower the cost of migration and mitigate some of the negative impacts. Reforming housing, pension, and health finance so that migrants and nonmigrants pay the same rates for these services would reduce inequality. Housing reforms would help prevent migrants from forming slum communities. Increasing access to health care would reduce the incidence of communicable disease. Of course, these reforms would increase the tendency for families to migrate. Consequently, the education system would need to be funded to accommodate growing numbers of students.

The hukou system has dampened migration, particularly family migration. The impact on individual migration is much less severe. Experiments with making the system more flexible, such as allowing migrants to apply for "blue" hukous (which give them temporary right of residence), should be accelerated. Streamlining the permit process by allowing migrants to receive all necessary permits from one source would reduce the costs and hardships migrants bear.

annex two

Survey of Literature on Inequality, Income Distribution, and Migration in China

Source	Summary	Abstract
Khan, A.R., and others. 1993. "Household Income and Its Distribution in China." In Keith Griffin and Zhao Renwei, eds., *The Distribution of Income in China.* New York: St. Martin's Press.	Describes rural, urban, and national income inequalities in 1988 and identifies contributing factors.	• Widens definition of income to include all disposable income (for example, payments in kind and agricultural output for self-consumption), making true per capita income in rural areas 39 percent higher and in urban areas 55 percent higher than State Statistical Bureau estimates. This adjustment implies that the urban bias of non-cash income raises rural-urban inequality and that household and national income are significantly underestimated. • Gini coefficients: rural inequality (Gini coefficient of 0.34) is significantly higher than urban inequality (0.23); total inequality (0.38) is higher than both rural and urban inequality because of large urban-rural inequality. • National inequality: the most important sources of income inequality are urban wages and in-kind subsidies to urban workers (especially

Source	Summary	Abstract
		housing subsidies), which are distributed relatively equally among the recipients but overall accrue to relatively rich urban workers, thus contributing 36 percent and 32 percent to national income inequality, respectively. • Rural income inequality: the most important factor explaining income inequality is the difference in income from production activities, which accounts for almost 75 percent of income and explains more than 60 percent of rural income inequality. The difference in wage income is also a significant contributor to rural inequality. • Urban income inequality: the two most important factors, wage inequality and differences in housing subsidies, contribute 34 percent and 24 percent, respectively, to urban inequality. Urban bias and regressive taxes and subsidies reduce rural incomes by 4 percent and raise urban income by 39 percent. If the benefits of subsidies and burden of taxes were neutral, rural incomes would be 23.5 percent higher and urban incomes would be 30.9 percent lower, and the rural-urban income ratio would increase from 41 percent to 74 percent. One weakness of this study is its interpretation of the contributing factors as "disequalizing" and "equalizing," rather than defining the effect of the factors depending on their overall contribution to inequality.
Renwei, Zhao. "Three Features of the Distribution of Income during the Transition to Reform." In Keith Griffin and Zhao Renwei, eds., *The Distribution of Income in China.* New York: St. Martin's Press.		China's reforms have resulted in a dualistic system that has three main characteristics: • The coexistence of a regulated state sector with relatively low levels of income inequality (Gini coefficient of 0.23) and a market-oriented sector characterized by high levels of income inequality (Gini of 0.49). • Distinct time dynamics: during 1978–84 inequality declined as the rural-urban income ratio rose from 42 percent to 54 percent, mainly because rural markets were opened, the commune system was dissolved, and rural (agricultural) terms of trade improved. During 1984–90 rural-urban income inequality increased because of declining productivity growth in agriculture, price and enterprise reforms in urban areas, and a decline in the rural terms of trade. • Urban nonwage payments rose rapidly after reforms were introduced. They now make up 30 percent of urban income and have a strongly disequalizing effect. • Recommends reducing inequality by deepening reforms and reducing nonwage sources of income to improve efficiency and incentives.

Source	Summary	Abstract
Khan, A. R. "The Determinants of Household Income in Rural China." In Keith Griffin and Zhao Renwei, eds., *The Distribution of Income in China.* New York: St. Martin's Press.	Describes rural income inequalities and identifies contributing factors.	• 74 percent of rural income comes from farm and nonfarm production activities; only 9 percent comes from wages and salaries. • Households that are located in privileged provinces have a higher stock of human capital (in terms of education), produce for the market rather than for self-consumption, are focused on nonfarm activities, and receive more income from production activities. • Variation in income among provinces is the most important factor explaining interregional differences. • An important source of rural income inequality results from wage employment, which is highly unequal (62 percent goes to the top 10 percent and 10 percent goes to the bottom 20 percent). Thus most rural income accrues to a small, privileged minority that lives in privileged regions, often close to urban centers, is likely to be a member of the Communist Party, and does not have a higher stock of human capital.
McKinley, Terry. 1996 *The Distribution of Wealth in Rural China.* New York: M.E. Sharpe.		• Wealth (Gini coefficient of 0.31) is distributed more equally than income (Gini of 0.34) in rural China. • Land accounts for 59 percent of rural wealth and housing, for 31 percent. • Land is distributed relatively equally (Gini coefficient of 0.31). Housing (Gini of 0.49) and other productive assets, such as financial assets, are not. • These characteristics are highly atypical for a developing country: in China the main source of rural income inequality is wage income rather than the return to assets such as land.
Riskin, Carl. "Income Distribution and Poverty in Rural China." In Keith Griffin and Zhao Renwei, eds., *The Distribution of Income in China.* New York: St. Martin's Press.	Analyzes rural poverty in China.	Using official Chinese conception of poverty plus adjustments for regional differences in price levels, finds that: • 105 million (12.7 percent) Chinese were below poverty line of 333 yuan in 1988. • Most poor people (64.5 percent) are not located in designated poverty regions. • Econometric analysis shows that the most powerful policy measures to address poverty in China are an increase in rural wage employment, an improvement of irrigation and drainage facilities in disadvantaged rural areas, and investment in human capital.

Source	Summary	Abstract
Knight, John and Song Lina. "Workers in China's Rural Industries." In Keith Griffin and Zhao Renwei, eds., *The Distribution of Income in China.* New York: St. Martin's Press.	Describes inequalities arising from rural wage employment.	• In 1988 rural industries (township, village, and private enterprises) employed 96 million people, or 24 percent of the rural labor force, and between 1978 and 1988 absorbed 62 percent of new entrants into the labor force. • Wage employment strongly exacerbates rural income inequalities because it benefits mainly the rich: richest 30 percent of rural population receives 84 percent of wage income. • Most rural enterprises are concentrated in richer and suburban rural areas. • Wage employment biased in favor of the educated, men, and members of the Communist Party.
Knight, John and Song Lina. "Why Urban Wages Differ in China." In Keith Griffin and Zhao Renwei, eds., *The Distribution of Income in China.* New York: St. Martin's Press.	Describes inequalities arising from urban wage employment.	• Urban wage distribution is highly equitable (Gini coefficient of 0.20), mainly because of administrative decisions rather than market forces. • Main disequalizing factor is noncash payments, such as housing and other services provided by enterprises and authorities, although urban income inequality is relatively equal for a developing country and more equal than rural income distribution. • Returns to education are low, even in the private sector. Discrimination exists on basis of sex and Communist party membership. • One striking factor of labor market is its high degree of job security, with job tenure basically guaranteed for life.
Knight, John and Li Shi. "The Determinants of Educational Attainment." In Keith Griffin and Zhao Renwei, eds., *The Distribution of Income in China.* New York: St. Martin's Press.	Analyzes distribution of education among Chinese population.	• Broad distribution and relatively equitable access to education, with a focus on primary and secondary education. • Limited supply of and access to higher education. • Women have 2.3 years less education than men. • Large urban-rural gap in educational attainment: urban people receive an average of 9.6 years of education—about 4.1 years more than rural people. • The strong link between provincial income and education levels in rural areas exacerbates education inequalities because rural areas generally have lower income levels.
Hussain, Athar, Peter Lanjouw, and Nicholas Stern. 1994. "Income Inequalities in China: Evidence from Household Survey Data." *World Development* 22(12): 1947–57.		• Income inequality in urban areas (Gini coefficients in 0.19–0.22 range) is lower than in rural provinces (Gini of 0.19–0.28). • Income inequalities at national level mainly result from intraprovincial inequalities. • Rural income: although farming income is the largest component of total income, nonfarming income is the largest contributor to inequality. • Urban income: wage income accounts for more than half of total income, but bonuses and irregular income (second jobs,

Source	Summary	Abstract
		commercial activities, hardship allowances) explain most urban inequality. • Authors conclude that with reforms and the move to market those components of income will rise most that contribute most to income inequalities—nonfarming activities and irregular income. There are, however, no empirical facts for this claim. Study excludes floating population of migrants and uses a narrow definition of income that ignores noncash income. Study argues that excluding migrants does not necessarily cause a downward bias because, although migrants have low-paid jobs, their families are mostly in rural areas, so their per capita income is not necessarily low. However, there is no empirical evidence for this claim.
Howes, Stephen, and Athar Hussain. 1994. "Regional Growth and Inequality in Rural China." London School of Economics.	Analyzes changes in and components of regional output inequalities. Uses State Statistical Bureau county-level output data for most of the 2,364 counties for 1985–91; broadest definition of rural is used, classifying 80 percent of the population as rural; uses constant 1980 consumer price index data.	• Rural output inequality rose during 1985–91; the main source was the rapid growth of the nonagricultural output of township and village enterprises. • Gini coefficient for total output per capita rose from 0.244 in 1985 to 0.335 in 1991 as a result of three main factors: 1. Net output increased by 6.4 percent a year, or 45 percent during 1985–91; nonagricultural output grew by 13 percent a year compared with agricultural output growth of 1.3 percent a year. 2. Gini coefficient for nonagricultural output (0.56 in 1991) was far higher than for agricultural output (0.23 in 1991). 3. The low rise in the Gini coefficient for nonagricultural output implies that township and village enterprise growth is not focused entirely on rich, eastern counties. The resulting contributions to the increase in overall output inequality are 11 percent for the increase in the Gini coefficient for nonagricultural output, 33 percent for the Gini coefficient for agricultural output, and 55 percent for the increase in the share of nonagricultural output. • Study disagrees with World Bank (1992) finding that poverty reduction stagnated or even reversed during late 1980s, and argues that output for almost all counties (except the poorest 5 percent) rose during 1985–90 and thus that poverty has continued to decline. It should be noted that output data and household data (such as those used in World Bank 1992) draw very different pictures. Output data estimates are higher due to the inclusion of retained rural output (for example, for enterprise consumption and investment, which does not go to households as income), the broader definition of rural areas (which includes towns where income and output per capita is generally higher), and

Source	Summary	Abstract
		measurement errors in data. The central question is how much retained rural output accrues to households in the form of, say, housing and social security benefits provided by workers' enterprises (noncash income) and how much is invested or consumed by enterprises without immediately benefiting workers. The actual figure is most likely between the two data estimates.
World Bank. 1995. "China: Regional Disparities." Report 14496-CHA.		• Disparities in output and especially consumption per capita between coastal and interior regions have increased since 1978 and especially since 1985. • Rural inequalities are larger than urban inequalities. • Disparities mainly result from competitive advantages of coastal areas stemming from advantages in transport, communications, and trade, which were exacerbated by market-oriented reforms since 1978, by policies that encourage economic activities in coastal regions (international commercial policies, fiscal and enterprise reforms, pilot programs) and attempt to curb migration flows, and by the lack of fiscal redistribution. • Regional social indicators (mortality rates, educational attainment, illiteracy) are broadly correlated with regional disparities in output and per capita income and consumption. • Based on findings for Henan and Sichuan, intraregional disparities are greater than interregional disparities. • Official data underestimate the size of the floating population living in richer coastal provinces, causing an overestimate of regional disparities. The floating population is likely to be around 10 percent of the coastal population; poverty is likely to be the most important force driving migration. • Study stresses the need to anticipate and manage migration flows as well as the need to reduce incentives for migration and its negative side effects by eliminating the urban bias of policies. • Study provides interesting discussion of data quality and related difficulties with analyzing disparities in China (appendix 2).
Jian, Tianlun, Jeffrey Sachs, and Andrew Warner. 1996. "Trends in Regional Inequality in China." NBER Working Paper 5412. National Bureau of Economic Research, Cambridge, Mass.	Analyzes convergence and divergence trends in GDP per capita among provinces and explains trends. Uses official Chinese data for 1952–93 for twenty-eight of thirty provinces using provincial price deflators.	• Study finds evidence for weak convergence for 1952–65, strong divergence for 1965–78, convergence for 1978–90, and divergence since 1990. • The two convergence measures are σ-convergence (standard deviation across regions of log real GDP per capita) and β-convergence (growth in per capita GDP relative to initial per capita GDP). • Initial convergence (1978–85) occurred mainly because of rapid growth in rural areas resulting form reforms. Later convergence (1985–90) occurred mainly because of continued

Source	Summary	Abstract
		growth in rural areas, especially areas near open coastal cities. Divergence since 1990 mainly due to rapid growth of coastal, urban areas, which grew by an average 7.4 percent a year faster than others. • Important finding that overall convergence during 1978–90 was mainly due to convergence among coastal provinces, while variance between coastal and interior regions remained constant in 1978–90 and then widened dramatically after 1990. • Study summarizes migration policies since 1950s and broad market reforms. • Predicts continued divergence; study points out that both labor and capital are flowing into richer, urban, coastal areas rather than poorer, rural, interior provinces. One of the study's weaknesses is that it excludes the floating population of 100–150 million people due to lack of data. Thus estimates are likely to be biased upward since actual GDP per capita is likely to be higher in rural areas and lower in urban areas (especially coastal provinces).
Yong, H.E., and Jean-Christophe Simon. 1995. "La distribution des revenus dans la transition economique de la Chine."	Criticizes official income measures that ignore other sources of income, such as corruption, rent seeking, and so on.	• Argues that income inequality derives not only from market forces but also from corruption, rent seeking, distortions created through the dual price system and its biased tax/subsidy system (for example, social benefits and unequal treatment in the privatization of housing), and so on. • Discusses shortcomings of official data that exclude these sources of income inequalities and thus are likely to underestimate true disparities. • Mentions individual cases of government distortions and corruption.
Ying, Yvonne. 1995. "Income, Poverty and Inequality in China during Transition." Research Paper 10. World Bank, Washington, D.C.	Uses State Statistical Bureau data: *China Statistical Yearbook* and Survey of Income and Expenditure of Urban Households in China.	• Makes similar points to above studies: income inequality and poverty fell between 1978 and 1984 and then stagnated or rose. • Summarizes agricultural policies since 1978 (household responsibility system and price policies). • Analyzes rural terms of trade: rise in agricultural prices during 1978–84 and then stagnation, mainly due to policy reforms; strong increase in fertilizer and pesticide prices since the late 1980s; strong rise in nonfarming output and income since the mid-1980s. • Concludes that urban inequality has stagnated since 1985: paper includes quantitative analysis which shows that the elimination or reduction of food subsidies caused urban inequalities to rise, while full employment policies and state enterprises' basic wage equalization eased income disparities somewhat.

Source	Summary	Abstract
		• Analyzing rural terms of trade has important implications for poverty analysis: it is not gross income that is most relevant, but disposable income (that is, prices) that are important—especially production costs (such as pesticide and fertilizer costs).
Deininger, Klaus, and Lyn Squire. 1996. "New Ways of Looking at Old Issues: Inequality and Growth." World Bank, Policy Research Department, Washington, D.C.		Using more comprehensive data, the authors yield interesting findings: • Initial inequality of assets (measured by land distribution) is more significant than income inequality in affecting subsequent growth. • Kuznets curve hypothesis does not hold for many countries: the most important factors affecting changes in inequality are policy-related and do not necessarily depend on a country's degree of development.
Guang Hua Wan. 1995. "Peasant Flood in China: Internal Migration and Its Policy Determinants." *Third World Quarterly* 16 (2).	Describes rural migration and migrant characteristics for 1986. Uses 1986 village survey for 230 villages in eleven regions (undertaken by the Chinese Academy of Social Sciences).	• In 1986, 37 percent of sample population were migrants. Of these, about 60 percent were intramigrants (moving within townships) and 40 percent were emigrants. • Only 4 percent of emigrants moved to big cities (that is, most emigrants moved to other rural areas). • Most migrants shifted into construction (23 percent) and industrial (34 percent) activities; less than 3 percent moved into agriculture. • Overall, migrants were not more educated: education deterred emigration but encouraged in-migration. The reason, according to the author, is that there is significant demand for less-skilled workers in rural towns and cities, while there is a significant demand for skilled and educated people in rural areas because of reforms. • Migration to urban areas is more likely to be seasonal, while migration to rural areas tends to be permanent. • Explanations: Urban reforms lagged behind rural reforms until mid-1980s, thus reducing the rural-urban gap in terms of income. Study argues that the dismantling of the commune system and the rapid growth of township, family, and cooperative enterprises explain a large share of intramigration flows. Study claims that the low share of rural-urban migration, its seasonal character, and the low female share of migrants is most likely explained by the lack or poor quality of social services for migrants in nonrural areas. • Policy recommendations: Government should provide migrants with social services in order to ease problems related to short-term migration (especially concerning infrastructure). It also should encourage urban-rural migration and lessen potential future pressure for rural-urban migration, and thus reduce the rural-urban gap, by eliminating privileges to urban

Source	Summary	Abstract
		residents (housing, education, health, and so on); reducing land fragmentation, which impedes agricultural productivity growth; and promoting rural urbanization to create jobs and reduce rural surplus labor. The study is for 1986; thus the dynamic of migration may have changed dramatically since then because the rural-urban gap has widened significantly and reforms in rural areas have slowed down. The study's finding of a negative relation between education and migration is questionable, and the finding that there is a flow of educated people from urban to rural areas is surprising. The survey's small sample may mean that villages in the sample are not representative, and there are a number of potential sources of measurement errors (for example, many permanent emigrants may not be included in study).
Li Debin. 1994. "The Characteristics of and Reasons for the Floating Population in Contemporary China." *Social Sciences in China* (winter): 65–72. Morf, Urs. 1994. "The Threat of Mass Migration in China." *Swiss Review of World Affairs* (April).	Describe characteristics and size of floating population using official State Statistical Bureau data for 1980–93.	• Floating population defined as mainly legal migrants—that is, those with permits or those who commute on a daily basis. This definition excludes/ignores those who migrate illegally. • In 1985 floating population was less than 10 million; in 1989 it was 60–80 million. • Most members of the floating population are farmers or agricultural workers who leave rural areas for urban areas to find nonagricultural jobs. • Causes of growth in floating population: increasing land fragmentation, growing population, and decline of cultivable land has created large surplus labor (about 200 million people in late 1980s); rapid growth of township and village enterprises and other private companies offer economic opportunities for workers; rapid growth of cities and attraction of cities for rural workers in terms of employment and living conditions; and policy biases favoring cities and special economic zones. • One of the studies argues that the slowdown in agricultural productivity growth can be partly explained by the migration flow of the most capable persons (that is, young men) from agricultural to nonagricultural jobs, leaving cultivation of the land to parents and other family members.
Sahota, Gian S. 1968. "An Economic Analysis of Internal Migration in Brazil." *Journal of Political Economy* 218–45.	Determines migrant characteristics and identifies migration flows among regions; uses household survey data (population census) for 1960.	• Develops a simple (and somewhat incomplete) model that defines migration as a function of average incomes within regions and education levels, population densities, geographic distances, and income dispersions across regions. • Provides interesting discussion of methodological issues; uses simultaneous equation model. • Basic findings are that migration is larger when education is higher in the destination area and lower in the origin area, although the effects are not large; that the earnings differential between origin and destination is the most relevant variable;

Source	Summary	Abstract
		that younger people are more sensitive to earnings differentials and thus are more likely to migrate; that urbanization and industrialization in the destination area are significant "pull" factors; that population density at destination was highly significant; and that income inequality was not important. Weaknesses: data refer only to a single year, people are considered migrants if they were born in a region other than their current location (even if they migrated many years ago and their current characteristics were acquired at that destination), and income data are highly aggregated.
Todaro, Michael P. 1976. *Internal Migration in Developing Countries.* Geneva: International Labour Organization.		• Excellent book on migration; compares approaches to migration (chapter 3), analyzes econometric approaches (chapter 4), and summarizes quantitative studies (chapter 5). • Chapter 3 outlines and makes important additions to the Harris-Todaro model.
Russel, Sharon Stanton. 1986. "Remittances from International Migration: A Review in Perspective." *World Development* 14(6): 677–96.	Reviews determinants and effects of remittances from international migration.	• Summarizes determinants and effects of remittances (especially tables 1 and 2 and figure 1). • Volume of remittances is large for many countries, totaled more than $23 billion in early 1980s. • Shows that most studies have found that remittances are used mainly for consumption and little for investment or saving. This pattern need not be unproductive, however, if remittances raise labor productivity. • Cites other studies that found equalizing effect of remittances and income distribution (in Mexico and Pakistan, for example).
Stark, Oded, J. Edward Taylor, and Shlomo Yitzhaki. 1986. "Migration, Remittances and Inequality: A Sensitivity Analysis Using the Extended Gini Index." Discussion Paper 23. Harvard University Migration and Development Program, Cambidge, Mass.	Uses alternative Gini indexes to show that the effect remittances have on inequality crucially depends on weight given in index to income of different groups. Uses village-level data for two Mexican villages near the U.S.-Mexican border.	• Finds that remittances from international migration to the United States reduce inequality. The effect of remittances from internal migration on inequality is ambiguous. • Analyzes the effect of using different weights in Gini index on inequality measure: stronger weight for poor households makes the effect of international remittances on inequality significantly less favorable. • Points out that low migration among the poor is mainly due to lack of resources to afford the journey.

Source	Summary	Abstract
Lipton, Michael. 1980. "Migration from Rural Areas of Poor Countries: The Impact on Rural Productivity and Income Distribution." *World Development* 8: 1–24.	Provides an overview of the conceptual links between migration, rural productivity, and income inequality in rural areas.	• Provides broad but sometimes confusing and contradictory review of conceptual links and empirical findings of migration effects. • Argues that emigration's effect on productivity can be ambiguous and depends on the response of the rural population to out-migration and the inflow of remittances. • Argues that emigration has a disequalizing effect—that is, it increases intra- and interrural inequalities, as well as rural-urban inequalities, for three reasons: 1. Migrant characteristics: most are young, male, and from poor or well-off households. Better-off, educated migrants receive higher payoffs from migration due to superior information, transferable skills, better networks, and better ability to pay for the initial costs of migration. Less-skilled households often lose productive family members without necessarily receiving significant returns, resulting in higher intrarural inequality. Villages near cities and rapidly growing areas benefit the most from migration, and migration from these areas is more likely and more profitable, resulting in higher interrural disparities. 2. Remittances: study argues that remittances mainly accrue to better-off families because they have a significant number of migrants with superior payoffs; that remittances are mainly used for purposes that do not benefit the poor; and that there is little evidence that remittances have a "trickle-down" effect. 3. Return migration: better-off people who have made accomplishments during emigration or have acquired skills that are useful in rural areas are the most likely to return; thus return migration tends to worsen both intra- and interrural disparities.
Adams, Richard H. 1992. "The Effects of Migration and Remittances on Inequality in Rural Pakistan." *The Pakistan Development Review* 31 (winter): 1189–1206.	Analyze effect remittances from internal and international migration have on rural income inequality.	• Studies compare predicted household income without migration with real household income with remittances from migration. The studies find that having an internal migrant within the household raises predicted household income by $6 a year while having an international migrant raises it by $25 a year (data are from Pakistan). • Remittances have a somewhat disequalizing effect on income inequality in four rural regions in Pakistan. • Reasons for neutral/disequalizing effects: migrants are mostly from rich or poor families, and volume of remittances is larger for richer households.

Source	Summary	Abstract
Adams, Richard H. 1989. "Worker Remittances and Inequality in Rural Egypt." *Economic Development and Cultural Change* 38: 45–71.		Data sets focus on a small and not representative sample of rural areas, studies ignore second- and third-order effects of remittances, and other studies find a more disequalizing effect of remittances in Pakistan because they use different data and a better methodology.
Mohtadi, Hamid. 1990. "Rural Inequality and Rural-Push versus Urban-Pull Migration: The Case of Iran, 1956–76." *World Development* 18(6): 837–44.	Analyzes contribution of urban pull and rural push factors to rural-urban migration between 1956 and 1976.	• Concludes that rural push factors were the main cause of heavy migration flows; specifically, the land reform of the period, which generated large inequalities by transforming a homogeneous society into two groups of landowners and landless wage laborers. • Confirms Lipton's (1980; see above) hypothesis that migration of the poor is mainly caused by rural push factors (either institutional factors, such as land distribution, or Malthusian factors), whereas rural-urban migration of richer, land-owning family members is mainly explained by urban pull factors (such as employment opportunities and higher wages and returns to education and skills in cities; an important pull factor was the oil boom of the late 1960s and the 1970s in Iran). One weakness of this study is that it uses a narrow definition of pull and push factors—employment and wage levels in origin and destination locations.
Mohtadi, Hamid. 1986. "Rural Stratification, Rural to Urban Migration, and Urban Inequality: Evidence from Iran." *World Development* 14(6): 713–25.	Analyzes effect rural-urban migration had on urban inequality in Iran between 1956 and 1976.	• Urban inequality increased in areas where a majority of migrants came from landless groups. Urban inequality decreased in areas where most migrants belonged to land-owning groups and families. • The main explanation for this finding is that the land-owning class has better education, kinship, and other essential resources (such as savings) to acquire better-paid jobs, usually in the formal sector, while landless groups usually had more trouble finding jobs and generally started off with informal sector jobs. One weakness of this study is that it measures inequality not as income inequality but uses the distribution of urban housing as a proxy for inequality.

Source	Summary	Abstract
Goldstein, Sidney, and Alice Goldstein. 1990. "China." In Charles B. Nam and others, eds., *International Handbook on Internal Migration.* Goldstein, Sidney, and Alice Goldstein. 1985. "Population Mobility in the People's Republic of China." Paper 95. East-West Population Institute, Honolulu.	Review studies of internal migration in China between 1949 and early 1980s.	• Review population policies and migration policies since 1949, emphasizing official policies as the driving force behind population movements. • Compare surveys and studies on migration in China prior to 1980. Most of these studies are based on official Chinese data, which strongly underestimate the true number of migrants. For example, official urban surveys estimate that there were just 10 million migrants in 1982; this number ignores the many temporary migrants who do not show up in official studies. • Describe administrative procedures and enforcement difficulties of migration.
Kam Wing Chan. 1994. "Urbanization and Rural-Urban Migration in China since 1982: A New Baseline." *Modern China* 20(3): 243–81.	Reviews the literature on migration and urbanization in China prior to 1992.	• Compares empirical studies and finds major problems concerning the definition of "urban": estimates of urban population for 1990 vary between 18.5 percent and 52.9 percent. • Urban in-migration accounted for about three-quarters of urban population growth in 1980s. In-migration peaked in 1987–88, then declined, then increased again in 1991–92. • About 65 percent of floating population made up of temporary migrants. • Another reason for rapid increase in size of urban population is reclassification of areas—that is, many regions were redefined as urban, increasing the relative and absolute size of the urban population.
Li, Wan Lang. 1992. "Migration, Urbanization and Regional Development: Toward a State Theory of Urban Growth in Mainland China." *Issues and Studies* (February): 84–102.		• Argues that migration flow in 1980s was much smaller than in 1950s and 1960s. • Argues that the main reasons for migration in the 1980s were family-related rather than job-related. One weakness of the study is that its argument is based on official data, which exclude many migrants (especially temporary migrants).
Goldstein, Sidney. 1990. "Urbanization in China, 1982–87: Effects of Migration and Reclassification." *Population and Development Review* 16 (4): 673–701.	Explains urbanization trends and analyzes factors contributing to urbanization in China for 1982–87; uses 1982 census and 1987 national survey to analyze urbanization and 1986 national migration survey to analyze internal migration.	• Focuses on effects of urban reclassification and migration. • Argues that the main factor behind strong growth in urban population was reclassification of urban areas in 1984, which reduced minimum population requirements and extended city and town boundaries to include adjoining areas and counties. • Reclassification, rather than migration, accounts for most of the dramatic increase in the labor force in towns (labor force grew by 231 percent) and cities (labor force grew by 33 percent). • Official data (1987 survey) identify only about 30 million

Source	Summary	Abstract
		migrants—that is, permanent migrants and migrants with permits. Thus it underestimates the true size and impact of migration by excluding most temporary and illegal migrants. • Most migration flows were intraprovincial (80 percent) rather than interprovincial (20 percent). • Most of the migrants included in the study were young and were moving to cities and especially to towns for nonagricultural jobs.
Yang, Xiushi, and Sidney Goldstein. 1990. "Population Movement in Zhejiang Province, China: The Impact of Government Policies." *International Migration Review* 23 (3).	Study based on official Chinese migration data for the region of Zhejiang.	• Finds evidence that migration in Zhejiang is mainly from rural to urban areas, is especially strong within the region (rather than interregional), is mostly temporary, and is mostly better-educated people who migrate from rural to urban areas. • Argues that government authorities allow temporary migration flows but prohibit permanent migration in order to reduce rural surplus labor, avoid excessive strain on urban infrastructure (because temporary migrants are not entitled to use all urban infrastructure and social services), and avoid having people leave rural areas and give up their land (thus reducing rural output and efficiency), and because temporary migration flows are possibly reversible, while permanent migration is not. • Urges the government to pay more attention to temporary migration because it strains urban infrastructure and leads to fundamental social and economic changes.
Bramall, Chris, and Marion E. Jones. 1993. "Rural Income Inequality in China since 1978." *The Journal of Peasant Studies* 21(1): 41–70.	Uses State Statistical Bureau data and 1984–85 data from the Rural Policy Research Unit of the State Council.	• Compares the two data sources and finds that changes in inequality in the official data are severely underestimated and that the true increase in inequality since 1978 has been far larger. • The alternative data source estimates a rural Gini coefficient of 0.40 in 1984—far higher than the official data. • The alternative data source finds that the rise in rural inequality mainly results from an increase in income inequality in the nonfarm sector. Moreover, rapid industrialization and unequal regional growth have resulted in higher interregional inequality. • Argues that migration is likely to exacerbate these inequalities because the experiences of other countries show that migrants are usually young and better skilled, hence worsening rural-urban disparities. One problem with this study is that it does not provide quantitative estimates of which factors have contributed to rising inequalities and by how much. Moreover, the claim that migration is likely to worsen rural-urban inequalities is not based on concrete evidence but rather on evidence from other countries, such as Brazil.

Source	Summary	Abstract
McKinley, Terry, and Keith Griffin. 1993. "The Distribution of Land in Rural China." *The Journal of Peasant Studies* 21(1): 71–84.	Uses data from a 1988 nationwide survey on landholdings of about 10,000 rural households.	• In 1988 rural land was distributed relatively equally in China. Moreover, the value of the land was highly equal by international standards. • Equal land distribution and land value has a minor effect on income inequalities because land has become an increasingly less important source of income. • The study also examines the effect of land leasing but finds that it has a very little disequalizing effect on rural incomes.
Cornia, Giovanni Andrea. 1994. "Income Distribution, Poverty and Welfare in Transitional Economies: A Comparison Between Eastern Europe and China." *Journal of International Development* 6(5): 569–607.		• Compares the experiences of transition economies in Eastern Europe (especially Romania and Russia) with China's experience since 1978. • Finds that poverty and inequalities have risen dramatically in Eastern Europe but have remained relatively constant/low in China. • Argues that macroeconomic reforms (trade liberalization, price reforms, and so on) are the main cause of rising poverty and disparities in Eastern Europe, and thus disputes the notion that the social system has failed the poor and benefits the better off. In fact, Cornia finds evidence that social systems in Eastern Europe had a significant equalizing and moderating effect. • Although overall poverty has been reduced during the initial phase of reform in China, the increase in income disparities has been induced mainly by internal policy measures such as declining social expenditures, fiscal decentralization, and dismantling of the commune system.

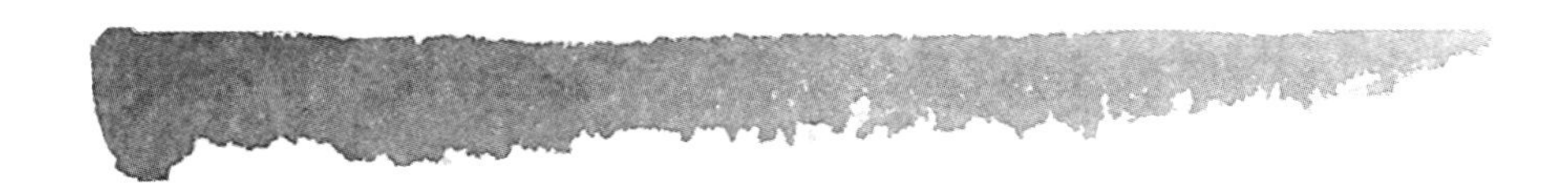

References

Background papers

Burgess, Robin. 1997. "Diversification and Welfare in Rural China." London School of Economics, Suntory-Toyota International Centre for Economics and Related Disciplines.

Burgess, Robin, and Mamta Murthi. 1996. "Land, Nutrition and Welfare in Rural China." London School of Economics, Suntory-Toyota International Centre for Economics and Related Disciplines.

Burgess, Robin, and Juzhong Zhuang. 1996. "Dimensions of Gender Bias in Intrahousehold Allocation in Rural China." London School of Economics, Suntory-Toyota International Centre for Economics and Related Disciplines.

Naga, Ramses Abdul, and Robin Burgess. 1996. "Determination and Prediction of Household Permanent Income." London School of Economics, Suntory-Toyota International Centre for Economics and Related Disciplines.

Ravallion, Martin, and Shaohua Chen. 1997. "When Economic Reform Is Faster than Statistical Reform: Measuring and Explaining Inequality in Rural China." World Bank, China and Mongolia Department, Washington, D.C.

Zhang, Tao, and Heng-fu Zou. 1996. "Determinants of Provincial Income Growth in China." World Bank, Policy Research Department, Washington, D.C.

Other references

ADB (Asian Development Bank). 1997. *Emerging Asia: Changes and Challenges*. Manila.

Ahuja, Vinod, Benu Bidani, Francisco Ferreira, and Michael Walton. 1997. *Everyone's Miracle? Revisiting Poverty Reduction and Inequality in East Asia*. A Directions in Development book. Washington, D.C.: World Bank.

Alesina, Alberto, and Dani Rodrik. 1994. "Distributive Policies and Economic Growth." *Quarterly Journal of Economics* 109 (May): 465–90.

Bardhan, Pranab. 1996. "Research on Poverty and Development Twenty Years after *Redistribution with Growth*." In Michael Bruno and Boris Pleskovic, eds., *Annual World Bank Conference on Development Economics 1995*. Washington, D.C.: World Bank.

Bauer, J., W. Feng, N.E. Riley, and X. Zhou. 1992. "Gender Inequality in Urban China: Education and Employment." *Modern China* 18 (3): 333–70.

Bramall, Chris, and Marion E. Jones. 1993. "Rural Income Inequality in China since 1978." *Journal of Peasant Studies* 21 (1): 41–70.

Burgess, Robin, Liwei Zhu, and Yun Ren. 1996. "Chinese Urban Household Expenditure Analysis 1986–1990." Economic and Social Research Council Series 14. London School of Economics and Political Science, Development Economics Research Programme.

Chen, Shaohua, and Martin Ravallion. 1996. "Data in Transition: Assessing Rural Living Standards in Southern China." *China Economic Review* 7 (1): 23–55.

China Ministry of Agriculture, Rural Economy Research Center Study Group. 1996a. "Analysis of the Income and Expenditure Patterns of Migrants in Shanghai." Beijing.

———. 1996b. "Zhongguo Nongcun Laodongli Liudong Yanjiu: Waichuzhe Yu Shuchudi." Summary report presented at the Ford Foundation's International Conference on China Rural Labor Mobility, June, Beijing.

China State Statistical Bureau. 1996. *China Statistical Yearbook 1996*. Beijing.

Cornia, Giovanni Andrea. 1994. "Income Distribution, Poverty and Welfare in Transitional Economies: A Comparison between Eastern Europe and China." *Journal of International Development* 6 (5): 569–607.

Datt, Gaurav, and Martin Ravallion. 1996. "Why Have Some Indian States Done Better than Others at Reducing Rural Poverty?" Policy Research Working Paper 1594. World Bank, Policy Research Department, Washington, D.C.

Deininger, Klaus, and Lyn Squire. 1996. "A New Data Set Measuring Income Inequality." *The World Bank Economic Review* 10 (3): 565–91.

Deininger, Klaus, Lyn Squire, and Tao Zhang. 1995. "Measuring Income Inequality: A New Data Base." World Bank, Policy Research Department, Washington, D.C.

Entwisle, Barbara, Gail E. Henderson, Susan E. Short, Jill Bouma, and Zhai Fengying. 1995. "Gender and Family Business in Rural China." *American Sociological Review* 60 (February): 36–57.

FBIS (Foreign Broadcast Information Service). 1995. "Rural Migrant Worker Economic Data Analyzed." Beijing.

———. 1996a. "PRC: Critique of Governor Responsibility System for Grain." Beijing.

———. 1996b. "PRC: Urban Floating Population Causes Social Problems." Beijing.

Feng, Lu. 1996. "Why Does China's Grain Policy Tend to Compromise the Efficiency Objective—A Neglected Issue for China's Grain Policy Research." Working Paper 1996002. Peking University, China Center for Economic Research, Beijing.

Findlay, Christopher. 1997. "Grain Sector Economic Reform in China." Paper presented at the Australian Agricultural and Resource Economics Society Conference, sponsored by the University of Adelaide's Chinese Economy Research Centre, January, Gold Coast.

Fishlow, Albert. 1996. "Inequality, Poverty, and Growth: Where Do We Stand?" In Michael Bruno and Boris Pleskovic, eds., *Annual World Bank Conference on Development Economics 1995*. Washington, D.C.: World Bank.

Gelbach, Jonah B., and Lant H. Pritchett. 1995. "Does More for the Poor Mean Less for the Poor? The Politics of Tagging." Policy Research Working Paper 1523. World Bank, Policy Research Department, Washington, D.C.

Griffin, Keith, and Ashwani Saith. 1981. "Growth and Equality in Rural China." International Labour Organization, Asian Employment Programme, Geneva.

Griffin, Keith, and Renwei Zhao, eds. 1993. *The Distribution of Income in China*. New York: St. Martin's Press.

Gundlach, Erich. 1996. "Solow Meets Market Socialism: Regional Convergence of Output per Worker in China." Kiel Working Paper 726. Kiel Institute of World Economics, Department IV, Germany.

Hare, Denise, and Shukai Zhao. 1996. "Labor Migration as a Rural Development Strategy: A View from the Migration Origin." Paper presented at the International Conference of Rural Labor in China, June 25–27, Beijing.

Howes, Stephen. 1993a. "Income Distribution: Measurement, Transition and Analysis of Urban China, 1981–1990." Ph.D. dissertation. London School of Economics and Political Science.

———. 1993b. "Income Inequality in Urban China in the 1980s: Levels, Trends and Determinants." Economic and Social Research Council Series 3. London School of Economics, Development Economics Research Programme.

Howes, Stephen, and Athar Hussain. 1994. "Regional Growth and Inequality in Rural China." London School of Economics, Suntory-Toyota International Centre for Economics and Related Disciplines.

Hussain, Athar. 1994. "Social Security in Present-Day China and its Reform." *American Economic Review Papers and Proceedings* 84 (May): 276–80.

Hussain, Athar, Peter Lanjouw, and Nicholas Stern. 1994. "Income Inequalities in China: Evidence from Household Survey Data." *World Development* 22 (12): 1947–57.

Jalan, Jyotsna, and Martin Ravallion. 1996a. "Are There Dynamic Gains from a Poor-Area Development Program?" Policy Research Working Paper 1695. World Bank, Policy Research Department, Washington, D.C.

———. 1996b. "Transient Poverty in Rural China." Policy Research Working Paper 1616. World Bank, Policy Research Department, Washington, D.C.

Jian, Tianlun, Jeffrey D. Sachs, and Andrew M. Warner. 1996. "Trends in Regional Inequality in China." NBER Working Paper 5412. National Bureau of Economic Research, Cambridge, Mass.

Khan, Azizur Rahman, Keith Griffin, Carl Riskin, and Zhao Renwei. 1992. "Household Income and Its Distribution in China." *The China Quarterly* 132: 1029–61.

Knight, John, and Li Shi. 1996. "Educational Attainment and the Rural-Urban Divide in China." *Oxford Bulletin of Economics and Statistics* 58 (1): 83–117.

Mason, Andrew D. 1997a. "Dimensions of the Labor Market in China: Rural Labor Markets and Rural, Urban, and Regional Linkages." Background paper prepared for this report. World Bank, Washington, D.C.

———. 1997b. "Gender Issues in the China Labor Market." Background paper prepared for this report. World Bank, Washington, D.C.

Maurer-Fazio, Margaret, Thomas G. Rawski, and Wei Zhang. 1997. "Gender Wage Gaps in China's Labor Market: Size, Structure, Trends." Paper presented at the annual meeting of the Asian Studies Association, March 13–17, Chicago.

McKinley, Terry. 1996. *The Distribution of Wealth in Rural China*. New York: M.E. Sharpe.

McKinley, Terry, and Keith Griffin. 1993. "The Distribution of Land in Rural China." *The Journal of Peasant Studies* 21 (1): 71–84.

Meng, Xin. 1996. "Regional Wage Gap, Information Flow and Rural-Urban Migration." Australian National University, School of Pacific and Asian Studies, Department of Economics Research, Canberra.

Meng, Xin, and Paul Miller. 1995. "Occupational Segregation and Its Impact on Gender Wage Discrimination in China's Rural Industrial Sector." *Oxford Economic Papers* 47 (1): 136–55.

Mody, Ashoka, and Fang-Yi Wang. 1995. "Explaining Industrial Growth in Coastal China: Economic Reforms...and What Else?" PSD Occasional Paper 2. World Bank, Private Sector Development Department, Washington, D.C.

Ravallion, Martin. 1997. "Poor Areas." In David Giles and Aman Ullah, eds., *The Handbook of Applied Economic Statistics.* New York: Marcel Dekkar.

Riley, Nancy E. 1995. "Chinese Women's Lives: Rhetoric and Reality." *Asia-Pacific Issues* 25. East-West Center, Honolulu.

Riskin, Carl. 1993. "Poverty in China's Countryside: Legacy and Change." In Pranab Bardhan, Mrinal Datta-Chaudhuri, and T.N. Krishnan, eds., *Development and Change: Essays in Honour of K.N. Raj.* Bombay: Oxford University Press.

Rozelle, Scott, Li Guo, Minggao Shen, and others. 1997. "Poverty, Networks, Institutions, or Education: Testing Among Competing Hypotheses on the Determinants of Migration in China." Paper presented at the annual meeting of the Association for Asian Studies, Chicago.

Summers, Robert, and Alan Heston. 1991. "The Penn World Table (Mark 5): An Expanded Set of International Comparisons, 1950–1988." *Quarterly Journal of Economics* 106: 327–68.

Taylor, J. Edward, and Irma Adelman. 1996. *Village Economies: The Design, Estimation and Use of Villagewide Economic Models.* Cambridge: Cambridge University Press.

Watson, Andrew, and Harry X. Wu. 1994. "Regional Disparities in Rural Enterprise Growth." In Christopher Findley, Andrew Watson, and Harry X. Wu, eds., *Rural Enterprises in China.* New York: St. Martin's Press.

West, Lorraine A.. 1995. "Regional Economic Variation and Basic Education in Rural China." Robert S. McNamara Fellowship Program. World Bank, Economic Development Institute, Washington, D.C.

World Bank. 1992. "China: Strategies for Reducing Poverty in the 1990s." Report 10409-CHA. China and Mongolia Department, Environment, Human Resources, and Urban Development Operations Division, Washington, D.C.

———. 1995a. *China: Macroeconomic Stability in a Decentralized Economy.* A World Bank Country Study. Washington, D.C.

———. 1995b. "China: Regional Disparities." Report 14499-CHA. China and Mongolia Department, Washington, D.C.

———. 1996a. "China: Higher Education Reform." Report 15573-CHA. Washington, D.C.

———. 1996b. "China: Issues and Options in Health Financing." Report 15278-CHA. China and Mongolia Department and Human Development Department, Washington, D.C.

———. 1996c. "China: Pension System Reform." Report 15121-CHA. China Resident Mission and China and Mongolia Department, Washington, D.C.

———. 1996d. *The Chinese Economy: Fighting Inflation, Deepening Reforms.* A World Bank Country Study. Washington, D.C.

———. 1996e. *Poverty Reduction and the World Bank: Progress and Challenges in the 1990s.* Washington, D.C.

———. 1996f. *World Development Report 1996: From Plan to Market.* New York: Oxford University Press.

———. 1997a. *At China's Table: Food Security Options.* Washington, D.C.

———. 1997b. "Memorandum and Recommendation of the President of the International Bank for Reconstruction and Development to the Executive Directors on a Proposed Loan of $30 Million to the People's Republic of China for a Qinba Mountains Poverty Reduction Project." Report P7090-CHA. Washington, D.C.

Wu, Guobao, Sue Richardson, and Peter Travers. 1995. "Rural Poverty and Its Causes in China." Working Paper 95. University of Adelaide, Chinese Economy Research Unit, Australia.

———. 1996. "Multiple Deprivation in Rural China." Working Paper. University of Adelaide, Chinese Economy Research Unit, Australia.

Wu, Harry, and Zhou Li. 1996. "Research on Rural to Urban Labour Migration in Post-Reform China: A Survey." Working Paper 71. University of Adelaide, Chinese Economy Research Unit, Australia.

Yang, Dennis Tao, and Hao Zhou. 1996. "Rural-Urban Disparity and Sectoral Labor Allocation in China." Paper presented at the annual meeting of the Association for Asian Studies, April, Honolulu, Hawaii.

Yang, L., and J.S. Zax. 1996. "Compensation for Holding up Half the Sky: Gender-linked Income Differences in Urban China" University of Colorado, Department of Economics, Boulder.